THE NATIONAL ACADEMIES
KECK FUTURES INITIATIVE

THE GENOMIC REVOLUTION

IMPLICATIONS FOR TREATMENT AND CONTROL OF INFECTIOUS DISEASE

WORKING GROUP SUMMARIES

Conference
Arnold and Mabel Beckman Center of the National Academies
Irvine, California
November 10-13, 2005

THE NATIONAL ACADEMIES PRESS
Washington, D.C.
www.nap.edu

THE NATIONAL ACADEMIES PRESS 500 Fifth Street, N.W. Washington, DC 20001

NOTICE: The working group summaries in this publication are based on working group discussions during the National Academies Keck *Futures Initiative* The Genomic Revolution: Implications for Treatment and Control of Infectious Disease Conference held at the Arnold and Mabel Beckman Center of the National Academies in Irvine, CA, November 10-13, 2005. The discussions in these groups were summarized by the authors and reviewed by the members of each working group. Any opinions, findings, conclusions, or recommendations expressed in this publication are those of the working groups and do not necessarily reflect the view of the organizations or agencies that provided support for this project.

Funding for the activity that led to this publication was provided by the W.M. Keck Foundation. Based in Los Angeles, the W.M. Keck Foundation was established in 1954 by the late W.M. Keck, founder of the Superior Oil Company. The Foundation's grant making is focused primarily on pioneering efforts in the areas of medical research, science, and engineering. The Foundation also maintains a Southern California Grant Program that provides support in the areas of civic and community services with a special emphasis on children. For more information, visit www.wmkeck.org.

International Standard Book Number 0-309-10109-3

Additional copies of this report are available from the National Academies Press, 500 Fifth Street, N.W., Lockbox 285, Washington, DC 20055; (800) 624-6242 or (202) 334-3313 (in the Washington metropolitan area); Internet, www.nap.edu.

Printed in the United States of America

THE NATIONAL ACADEMIES

Advisers to the Nation on Science, Engineering, and Medicine

The **National Academy of Sciences** is a private, nonprofit, self-perpetuating society of distinguished scholars engaged in scientific and engineering research, dedicated to the furtherance of science and technology and to their use for the general welfare. Upon the authority of the charter granted to it by the Congress in 1863, the Academy has a mandate that requires it to advise the federal government on scientific and technical matters. Dr. Ralph J. Cicerone is president of the National Academy of Sciences.

The **National Academy of Engineering** was established in 1964, under the charter of the National Academy of Sciences, as a parallel organization of outstanding engineers. It is autonomous in its administration and in the selection of its members, sharing with the National Academy of Sciences the responsibility for advising the federal government. The National Academy of Engineering also sponsors engineering programs aimed at meeting national needs, encourages education and research, and recognizes the superior achievements of engineers. Dr. Wm. A. Wulf is president of the National Academy of Engineering.

The **Institute of Medicine** was established in 1970 by the National Academy of Sciences to secure the services of eminent members of appropriate professions in the examination of policy matters pertaining to the health of the public. The Institute acts under the responsibility given to the National Academy of Sciences by its congressional charter to be an adviser to the federal government and, upon its own initiative, to identify issues of medical care, research, and education. Dr. Harvey V. Fineberg is president of the Institute of Medicine.

The **National Research Council** was organized by the National Academy of Sciences in 1916 to associate the broad community of science and technology with the Academy's purposes of furthering knowledge and advising the federal government. Functioning in accordance with general policies determined by the Academy, the Council has become the principal operating agency of both the National Academy of Sciences and the National Academy of Engineering in providing services to the government, the public, and the scientific and engineering communities. The Council is administered jointly by both Academies and the Institute of Medicine. Dr. Ralph J. Cicerone and Dr. Wm. A. Wulf are chair and vice chair, respectively, of the National Research Council.

www.national-academies.org

THE NATIONAL ACADEMIES KECK *FUTURES INITIATIVE* GENOMICS STEERING COMMITTEE

ROBERT H. WATERSTON (Chair) (NAS/IOM),* William H. Gates III Endowed Chair in Biomedical Sciences, Chair and Professor, Department of Genome Sciences, University of Washington

ARUP K. CHAKRABORTY (NAE), Warren and Katherine Schlinger Distinguished Professor and Chairman, Department of Chemical Engineering, University of California, Berkeley

RONALD W. DAVIS (NAS), Professor of Biochemistry and Genetics, Director, Stanford Genome Technology Center, Stanford University School of Medicine

HELENE D. GAYLE, M.D. (IOM), Director of HIV, TB and Reproductive Health, Bill & Melinda Gates Foundation

DAVID GINSBURG, M.D. (IOM),** Investigator, Howard Hughes Medical Institute, James V. Neel Distinguished University Professor, Departments of Internal Medicine and Human Genetics, University of Michigan Medical School

RICHARD M. KARP (NAS/NAE), University Professor, University of California, Berkeley, Senior Research Scientist, International Computer Science Institute

DIANE MATHIS (NAS), William T. Young Chair in Diabetes Research, Professor of Medicine and Co-Head, Section on Immunology and Immunogenetics, Joslin Diabetes Center, Harvard Medical School

---

* Also Chair, Planning Committee.

** Also Member, Planning Committee.

THE NATIONAL ACADEMIES KECK *FUTURES INITIATIVE* GENOMICS PLANNING COMMITTEE

BRUCE BEUTLER, M.D., Professor, Department of Immunology, The Scripps Research Institute
SHU CHIEN (NAS/NAE/IOM), University Professor of Bioengineering and Medicine, Chair, Department of Bioengineering, University of California, San Diego
VANESSA NORTHINGTON GAMBLE, M.D. (IOM), Director of the National Center for Bioethics in Research and Health Care, Tuskegee University
ALAN E. GUTTMACHER, M.D., Deputy Director of the National Human Genome Research Institute, National Institutes of Health
DAVID HAUSSLER, Director of the Center for Biomolecular Science & Engineering, Professor of Computer Science, Howard Hughes Medical Institute, University of California
MUIN J. KHOURY, M.D., Director of the Office of Genomics and Disease Prevention, Centers for Disease Control and Prevention
LEONID KRUGLYAK, Professor of Ecology and Evolutionary Biology and the Lewis-Sigler Institute for Integrative Genomics, Princeton University
RICK LIFTON, M.D. (NAS/IOM), Chairman of the Department of Genetics, Professor of Medicine, Genetics, Molecular Biophysics and Biochemistry, Howard Hughes Medical Institute, Yale University School of Medicine
DAVID LOCKHART, President and CSO (Co-founder and Director), Ambit Biosciences
DEIRDRE MELDRUM, Director of the NIH Center for Excellence in Genomic Sciences (CEGS) Microscale Life Sciences Center, Professor of Electrical Engineering, University of Washington
PAUL SCHAUDIES, Assistant Vice President and Division Manager, Biological and Chemical Defense Division, Science Applications International Corporation

Staff

KENNETH R. FULTON, Executive Director
MARTY PERREAULT, Program Director
MEGAN ATKINSON, Senior Program Specialist
RACHEL LESINSKI, Senior Program Specialist
ALEX COHEN, Research Associate / Programmer

The National Academies Keck *Futures Initiative*

The National Academies Keck *Futures Initiative* was launched in 2003 to stimulate new modes of scientific inquiry and break down the conceptual and institutional barriers to interdisciplinary research. The National Academies and the W.M. Keck Foundation believe that considerable scientific progress will be achieved by providing a counterbalance to the tendency to isolate research within academic fields. The *Futures Initiative* is designed to enable scientists from different disciplines to focus on new questions, upon which they can base entirely new research, and to encourage and reward outstanding communication between scientists as well as between the scientific enterprise and the public.

The *Futures Initiative* includes three main components:

Futures Conferences

The *Futures* Conferences bring together some of the nation's best and brightest researchers from academic, industrial, and government laboratories to explore and discover interdisciplinary connections in important areas of cutting-edge research. Each year, some 100 outstanding researchers are invited to discuss ideas related to a single cross-disciplinary theme. Participants gain not only a wider perspective but also, in many instances, new insights and techniques that might be applied in their own work. Additional pre- or postconference meetings build on each theme to foster further communication of ideas.

Selection of each year's theme is based on assessments of where the intersection of science, engineering, and medical research has the greatest potential to spark discovery. The first conference explored *Signals, Decisions, and Meaning in Biology, Chemistry, Physics, and Engineering.* The 2004 conference focused on *Designing Nanostructures at the Interface between Biomedical and Physical Systems.* The theme of the 2005 conference was *The Genomic Revolution: Implications for Treatment and Control of Infectious Disease.* In 2006, the conference will focus on *Smart Prosthetics: Exploring Assistive Devices for the Body and Mind.*

Futures Grants

The *Futures* Grants provide seed funding to *Futures* Conference participants, on a competitive basis, to enable them to pursue important new ideas and connections stimulated by the conferences. These grants fill a critical missing link between bold new ideas and major federal funding programs, which do not currently offer seed grants in new areas that are considered risky or exotic. These grants enable researchers to start developing a line of inquiry by supporting the recruitment of students and postdoctoral fellows, the purchase of equipment, and the acquisition of preliminary data—which in turn can position the researchers to compete for larger awards from other public and private sources.

National Academies Communication Awards

The Communication Awards are designed to recognize, promote, and encourage effective communication of science, engineering, medicine, and interdisciplinary work within and beyond the scientific community. Each year the *Futures Initiative* honors and rewards individuals with three $20,000 prizes, presented to individuals who have advanced the public's understanding and appreciation of science, engineering, and/or medicine. Awards are given in three categories: book author; newspaper, magazine, or online journalist; and TV/radio correspondent or producer. The winners are honored during the *Futures* Conference.

Facilitating Interdisciplinary Research Study

During the first 18 months of the Keck *Futures Initiative,* the Academies undertook a study on facilitating interdisciplinary research. The study examined the current scope of interdisciplinary efforts and provided recommendations as to how such research can be facilitated by funding organizations and academic institutions. *Facilitating Interdisciplinary Research* (2005) is available from the National Academies Press (www.nap.edu).

About the National Academies

The National Academies comprise the National Academy of Sciences, the National Academy of Engineering, the Institute of Medicine, and the National Research Council, which perform an unparalleled public service by bringing together experts in all areas of science and technology, who serve as volunteers to address critical national issues and offer unbiased advice to the federal government and the public. For more information, visit www.national-academies.org.

About the W.M. Keck Foundation

Based in Los Angeles, the W.M. Keck Foundation was established in 1954 by the late W.M. Keck, founder of the Superior Oil Company. The Foundation's grant making is focused primarily on pioneering efforts in the areas of medical research, science, and engineering. The Foundation also maintains a Southern California Grant Program that provides support in the areas of civic and community services with a special emphasis on children. For more information, visit www.wmkeck.org.

The National Academies Keck *Futures Initiative*
5251 California Avenue – Suite 230
Irvine, CA 92617
949-387-2464 (Phone)
949-387-0500 (Fax)
www.keckfutures.org

Preface

At the National Academies Keck *Futures Initiative* The Genomic Revolution: Implications for Treatment and Control of Infectious Disease conference, participants were divided into interdisciplinary working groups. The groups spent eight hours over four days exploring diverse challenges at the interface between science, engineering, and medicine.

The goals of the working groups were to spur new thinking, to have people from different disciplines interact, and to forge new scientific contacts across disciplines. The working groups were not expected to solve the particular problems posed to the group, but rather to come up with a consensus method of attack and a thoughtful list of what we know and don't know how to do, and what's needed to get there. The composition of the groups were intentionally diverse, to encourage the generation of new approaches by combining a range of different types of contributions. The groups included researchers from science, engineering, and medicine, as well as representatives from private and public funding agencies, universities, businesses, journals, and the science media. Researchers represented a wide range of experience—from postdoc to those well-established in their careers—from a variety of disciplines that included genetics, microbiology, immunology, bioengineering, electrical engineering, chemistry, ecology, mechanical engineering, philosophy/ethics, law, medicine, epidemiology, and public health.

The conference committee had five objectives for the working groups:

- To approach the application of genomics to infectious disease from the perspective of problems having potentially revolutionary impact, rather than from the perspective of extensions of existing technology;

- To allow a group of people with a broad range of backgrounds to pool their insights and creativity to work on a shared interesting problem;
- To identify ideas and insights common to a number of working groups, and to identify how advances in genomics and their application might have a very large impact on the treatment and control of infectious disease;
- To identify the best (by whatever metrics seem to fit) big problems in the treatment and control of infectious disease to which genomics might be applied, and to identify gaps in knowledge that limit progress in the solution of these problems; and
- To allow individuals to make connections with one another in small working groups.

The groups needed to address the challenge of communicating and working together from a diversity of expertise and perspectives, as they attempted to solve a complicated, interdisciplinary problem in a relatively short time. Each group decided on its own structure and approach to tackle the problem. Some groups decided to refine or redefine their problems, based on their experience.

Each group presented two brief reports to the whole conference: (1) an interim report on Friday to debrief on how things were going, along with any special requests (such as an expert in DNA sequencing to talk with the group); and (2) a final briefing on Sunday where each group:

- Provided a concise statement of the problem
- Outlined a structure for its solution
- Identified the most important gaps in science and technology and recommended research areas needed to attack the problem
- Indicated the benefits to society if the problem could be solved

Each working group included a graduate student in a university science writing program. Based on the group interaction and the final briefings, the students wrote the following summaries, which were reviewed by the group members. These summaries describe the problem and outline the approach taken, including what research needs to be done to understand the fundamental science behind the challenge, the proposed plan for engineering the application, the reasoning that went into it, and the benefits to society of the problem solution.

Contents

Diagnosis

Natural Variation

APPENDIXES

Conference Summary

THE NATIONAL ACADEMIES KECK *FUTURES INITIATIVE* LAUNCHES A GENOMIC ATTACK ON INFECTIOUS DISEASE

By Kiryn Haslinger

One of the biggest questions on everyone's mind each winter is how to avoid getting the flu. Influenza and its fellow infectious diseases—which range from moderately troublesome bugs such as the common cold to lethal assassins like HIV—are collectively the most potent source of illness and mortality across the world. Just as understanding the criminal mind may be the most effective way to deter human villains, a similarly penetrating approach could be applied against these microbial bad guys.

This broad imperative provided the impetus to bring together 150 researchers, policy makers, foundation representatives, and members of the science media during November 10-13, 2005, at a conference to discuss solutions to the growing problem of infectious diseases using the field of genomics. The third annual conference of the National Academies Keck *Futures Initiative* (NAKFI), "The Genomic Revolution: Implications for Treatment and Control of Infectious Disease," invited its participants to develop creative ways to attack dangerous microbes through understanding their fundamental genomic compositions.

The Rise of Genomes

At the intersection between advances in molecular biology and genetics and advances in computing, techniques have been developed to rapidly and efficiently read genome sequences. This research area, which applies computational methods to large-scale genetic analysis, is known as "genomics."

Genomics is a tool scientists, engineers, and medical clinicians use to gain a detailed understanding of the biological roots of health and disease. A genome is the collection of chromosomes in an organism, which carries all its genes. It is the instruction book for creating a paramecium, pea plant, or person—written in DNA. Possessing the recipes for the Earth's creatures is extremely valuable. Knowing these chemical codes is a clue to figuring out how organisms function.

Nearly every human ailment is genetically based, at least to some degree, and knowing the ins and outs of genomes will be vital for the next phase of medical diagnosis and treatment of a wide variety of diseases. The goal of the third annual NAKFI conference was to bring to the table bold, innovative ideas about harnessing genomic science to mitigate the spread of infectious disease.

Facilitating Collaboration

The focus of the *Futures Initiative* is interdisciplinary research. "Discovery comes at the interstices of disciplines," said National Academy of Engineering President Wm. A. Wulf as he kicked off the four-day conference. Genomics is itself an interdisciplinary field, the progeny discipline born from the marriage of genetics and computer science. A primary objective of the meeting was to present a fertile ground for additional cross-disciplinary collaborations: stimulating alliances between biochemistry and engineering, immunology and economics, pharmacology and politics, science and communication. "I hope that by the end of this session," said Dr. Wulf, "you will establish lifelong relationships with people from fields you don't normally interact with."

To foster such relationships, Dr. Wulf introduced a series of tutorial sessions intended to explain the state of the science of various specialties, so that researchers could communicate clearly with one another across disciplines. In one tutorial, Gary Nabel, director of the Vaccine Research Center at the National Institute of Allergy and Infectious Diseases, discussed genomics, structural biology, and rational vaccine design. Using genomics, he asked, "How can we create new paradigms to create highly effective vaccines?" He stressed that the key would be gaining a better understanding of gene function and evolution.

In another session, to emphasize the dynamic interplay between humans and microbes Stanford University Professor David Relman declared, "We are 10 parts microbial and one part human," as though he were a bartender mixing up a biological cocktail. He then led conference participants on a tour of disease epidemiology, highlighting the importance of studying the genomes of pathogens as well as that of the human host, a point that was continually revisited throughout the conference.

Michael Waterman, a professor at the University of Southern California, in his talk attempted to demystify the technical work involved in computational biology and bioinformatics. These highly mathematical specialties form the core of genomics, and it is becoming increasingly vital for researchers to learn about automated sequencing, microarray technology, and optical mapping.

Interspersed with other technical tutorials on topics such as human genetic variation were talks about the societal impact of infectious disease. Austin Demby, a senior staff fellow at the Global AIDS Program of the Centers for Disease Control and Prevention, discussed the needs of developing countries and the unique delivery and implementation issues that face parts of the world most affected by infectious disease, such as sub-Saharan Africa.

Drilling for Oil in Orange County

The tutorials served as a stepping-off point for the participants to address concrete problems in small working groups. Each of 11 groups was presented with an outstanding challenge in the area of genomic approaches to infectious disease and asked to develop a scientific plan to address it. As conference chair, Robert Waterston, a professor at the University of Washington and head of one of the genome-sequencing centers that led the Human Genome Project, announced his charge to the working groups, and emphasized the value of problem-oriented thinking. Dr. Waterston also encouraged the group members to connect with researchers in other disciplines, joking that the real challenge of the working groups would be to figure out "how to have fun locked in a room with 10 of your newest colleagues."

Throughout eight hours of intense discussion spread over the four-day meeting, conference participants brainstormed about potential plans for designing technology to improve rapid response to disease, developing an inexpensive diagnostic test for pathogens, preventing the next pandemic flu, creating a device to detect and identify pathogens, and sequencing an individual's genome for under $1,000. Other groups focused on such vital topics as determining the role of public health in integrating genomics into disease control and devising new therapies by harnessing natural genetic variation in disease resistance.

The groups were not expected to solve these pressing scientific conundrums during the conference, but rather they were asked to assess the related advances that already exist and identify the gaps that must be filled. Their primary objective was to outline a method of attack for their problem. Richard Foster, a board member of the W. M. Keck Foundation, which funds the $40 million 15-year NAKFI program, spoke to the assembled groups about the life and spirit of W. M. Keck, an oil pioneer who drilled 23 dry holes before he hit his first gusher. "We expect you all to drill dry holes," he stressed. "We want you to get in there and start drilling right away. That's what this meeting is all about."

Drill the groups did, intensely digging into their individual areas of expertise to drive discussions about potential solutions to the problems presented. Each group, complete with strong personalities that ranged from open-minded and optimistic to imperious and despairing, was a microcosm for real-world research groups who work everyday to explore possible solutions to substantial scientific problems. Group members debated how best to phrase the questions they would set out to answer and then traded expertise in hopes of generating innovative ideas. At the end of the meeting, each group reported to the others what they had come up with.

One group outlined a detailed framework for governments to react rapidly to emerging disease, providing details for the key stages of surveying and monitoring pathogens, identifying infectious agents, and treating infected individuals. Another group developed blueprints for a device to monitor the environment (either air or biological fluids like mucus, where infectious agents may reside) by rapidly sequencing microbial genomes. Using a combination of microarray methods and nanotechnology, the group conceptually designed a disposable chip to search for malaria or HIV by sequencing DNA. A third group also incorporated nanoscience in their plan to engineer a laboratory incubator for culturing and crystallizing microbes to expedite vaccine research. Still another group devised a detailed budget for dispersing $100 million to prevent a pandemic flu, theoretically allotting $50 million to create an international flu research center and the remaining funds for supporting competitive grants for vaccine and antiviral research.

These working groups provided a fertile environment for communication among scientists, engineers, and medical researchers, many of whom discovered a valuable opportunity for interdisciplinary collaboration. The *Futures Initiative* aims to spark such relationships and offers a rewarding incentive for researchers to do so: up to $75,000 to fund innovative research and continue collaborative dialogues that emerged from the conference. The initiative supplies $1 million annually for such seed grants, awarded competitively to conference participants.

Communicating Science

Another major goal of the initiative is to encourage communication of scientific discoveries and ideas to the public. Effective communication, particularly on the topic of genomics, not only educates the public about scientific progress but also empowers individuals to make informed decisions about their own health. If you understand the basics of genomics and your genome indicates that you are predisposed to, say, high blood pressure, you know that it is particularly important to avoid cigarettes and limit the amount of salt in your diet. Genomic sequencing can also tell you if you have been exposed to a disease like malaria, which can help you expedite treatment and prevent spreading it to others. The working groups that were charged with outlining the role of public health in integrating genomics came to the realization that clear communication about this subject is itself a weapon against infectious disease.

Clear, accessible writing and broadcasting can serve to interest and excite people about genomic discovery and to dilute the mainstream sensationalism that asserts the utility of the field lies in ethically questionable applications like designer babies and cloned pets. A core feature of the *Futures Initiative* is the presentation of the National Academies Communication Awards, given to an author, a journalist, and a television or film producer. The $20,000 awards recognize excellence in reporting and communicating science, engineering, and medicine to the general public, and the winners were selected from more than 200 entries. At the conference, one award was given to John M. Barry for his book *The Great Influenza: The Epic Story of the Deadliest Plague in History.* Gareth Cook, a journalist for *The Boston Globe*, was presented an award for his coverage of the national debate on stem cells. An additional prize was given to Thomas Levenson for his television program on the evolution of life in the cosmos, WGBH NOVA's "Origins: Back to the Beginning."

By inviting graduate student science writers to attend its meetings and write articles describing the progress of the working groups, NAKFI encourages young writers to communicate science effectively to a diverse audience. Eleven students, selected from universities from around the country, had the opportunity at November's conference to attend the same tutorials as the researchers and participate in the working groups. Their articles have been collected here to create a comprehensive summary of the working groups' conclusions.

The Revolution Continues

The devastation wreaked by infectious disease spurred the conference participants to have valuable discussions and consider novel genomic solutions to controlling disease. In an era of fast-mutating deadly viruses like SARS and avian flu, and persistent killers like malaria, tuberculosis, and HIV, there is certainly a compelling reason to focus modern developments in genomics while trying to treat infectious disease. Throughout the conference, discoveries about genomics and infectious disease permeated the Arnold and Mabel Beckman Center, even when the working groups were on hiatus. Poster sessions were held each afternoon, allowing researchers to present relevant unpublished work from their laboratories and providing ample opportunity for networking. In several cases, meeting participants took inspiration from these informal sessions back with them to the working groups.

Perhaps because of the striking relevance of the problems at hand, there was a great deal of overlap among the discussions by various groups and consensus on recommendations for future research. The overarching theme of the conference was the need for individualized approaches to medicine, which, because of genomic variation, will be the vital next step in advancing medical treatment.

At the end of the four days, working group members announced their progress and recommendations for future research (see the write-ups that follow for specific summaries). Some recommendations were made by multiple groups. For instance, several groups projected that rapid, inexpensive sequencing would dramatically boost the progress toward their goals. Such a revolution in sequencing would advance the capability of diagnostic tests, enable efficient environmental monitoring for pathogens, and spur new technologies that we cannot even envision at present. The group responsible for outlining a path to the $1,000 genome vigorously debated the ethical issues that this cheap technology would introduce, while all the other groups took the value of such technology for granted.

There are risks, technical and ethical, inherent in pursuing this type of research. In a tutorial on conducting team science, Mary E. Lindstrom, vice-provost of research at the University of Washington, warned, "If you're going to take risks, you cannot expect 100 percent success."

The *Futures Initiative* has made valuable investments in scientific risk taking since it was launched in 2003, lauding and supporting bold efforts in both scientific research and communication. And the payoff may be very high: helping to mitigate the ravages of infectious disease.

Identify What Technological Advances in the Fields of Science and Engineering Need to Be Developed (Either New Technology or Novel Integration of Existing Technologies) to Improve Rapid Response to New or Emerging Diseases

WORKING GROUP DESCRIPTION

Background

Only a small fraction of the microbial life on Earth has been actively studied and characterized. Extremophiles have demonstrated the ability of life to exist in environments that appear harsh by existing standards. These adaptations clearly demonstrate the ability of bacteria and viruses to make genomic modifications that result in novel functions. As the rapid movement of people increases around the globe, the probability of becoming infected with or transmitting a previously undiscovered infectious organism is constantly increasing. When one compounds this increased mobility with the naturally changing genomes of existing bacterial and viral pathogens, the result is a precarious balance between pandemic and small isolated outbreaks.

Genetically modified organisms are already in use commercially to enhance milk production as well as to make crops resistant to insects and viruses. Some of these modifications are achieved by selective breeding and others by deliberate genetic modifications. Is it possible to safely create modified organisms that prey on specific target organisms for therapeutic or diagnostic purposes?

The Problem

Using the collective wisdom represented by this team, refine the working group topic into discrete tasks and outline steps required to approach each of the steps.

Initial References

Brown, K. 2003. Innovations: working weeds. *Scientific American* April 2003.

Freeland, S. J., and L. D. Hurst. 2004. Evolution encoded. *Scientific American* April 2004.

Gibbs, W. W. 2004. Synthetic life. *Scientific American* May 2004.

Jones, S. M., H. Feldmann, U. Stroher, J. B. Geisbert, L. Fernando, A. Grolla, H. D. Klenk, N. J. Sullivan, V. E. Volchokov, E. A. Fritz, K. M. Daddario, L. E. Hensley, P. B. Jahrling, and T. W. Geisbert. 2005. Live attenuated recombinant vaccine protects nonhuman primates against Ebola and Marburg viruses. Nature Medicine 11(7): 786-790.

WORKING GROUP SUMMARY

Summary written by:

Amos Kenigsberg, Graduate Student, Science Journalism, Boston University

Focus group members:

- Katie Brenner, Doctoral Candidate, Bioengineering, California Institute of Technology
- Frederic Bushman, Professor, Microbiology, University of Pennsylvania
- Amos Kenigsberg, Graduate Student, Science Journalism, Boston University
- Mary E. Lidstrom, Associate Dean, New Initiatives in Engineering, University of Washington
- Ulrich Melcher, R. J. Sirny Professor, Biochemistry and Molecular Biology, Oklahoma State University
- Gregory A. Petsko, Gyula and Katica Tauber Professor of Biochemistry and Chemistry, and Director, Rosenstiel Center, Rosenstiel Basic Medical Sciences Research Center, Brandeis University

- Alan Porter, Technology Policy and Assessment Center, Georgia Institute of Technology, and Evaluation Consultant, National Academies Keck *Futures Initiative*
- R. Paul Schaudies, Assistant Vice-President, Biological and Chemical Defense Division, Science Applications International Corporation
- Upinder Singh, Assistant Professor, Internal Medicine, Microbiology and Immunology, Stanford University
- Luis Villarreal, Professor, Molecular Biology and Biochemistry, University of California, Irvine

Summary

At first glance this working group's charge seemed straightforward: identify some technologies that will help us identify and treat new diseases as they are born or begin harming people. But during our first discussion, it became clear that we had a bigger goal in mind. It seemed to us that in order to address this question, we had to understand—and improve—the entire approach to how we fight emerging disease. Rapid response, our discussions emphasized, involves a broad range of institutions and technologies, from engineering bacteriophages to smart toilets to the marketability of vaccines. And it all seemed so simple—at first.

Our group decided to propose not just technologies but also a framework for a comprehensive response to emerging diseases—whether entirely new or resurfacing. This framework focuses on four sequential stages that compose the strictly technical aspect of rapid response. We sought to improve the effectiveness of the science in each of the four stages such that we could detect new and emerging diseases earlier and respond to them more quickly and effectively. Imagine, for example, that if as soon as the flu of 1918 evolved, scientists identified it immediately in its natural reservoir and stopped it before it harmed a single person. Our group believes this framework can bring us closer to that level of effective rapid response. Here are the four stages of rapid response:

1. Surveillance and monitoring for infectious agents
2. Detection of disease state and identification of causative agent
3. Initial response
4. Later response

Our recommendations to improve each stage:

Surveillance and monitoring: Detecting and identifying potential infectious agents before they spread widely in the human population

The most rapid part of rapid response comes from identifying a problem quickly by using good methods of surveillance. Our group saw two major types of early pathogen detection: biological monitoring and information analysis.

Biological monitoring is a system designed to detect and identify pathogens anywhere in the environment, such as in animal populations, on a public bus, at the scene of a bioterror attack, or in the body of an unknowingly infected person. The group thought that one important way to detect infectious agents this early in the process is through devices that provide constant, widespread environmental monitoring for microbes. These background monitors should focus on some key areas of high risk and high-population density, like hospitals and public transportation. One mechanism to do this would be to install ventilation systems that sample and analyze air as it circulates. In the future these environmental monitors could spread to other locations; surfaces in public buildings, like doorknobs or handrails, could also be surveilled for pathogens. In addition to looking for organisms themselves, environmental monitoring systems can look for early effects of infections in people. For example, during the SARS outbreak, infrared sensors were used to measure the temperature of people's foreheads and faces to see if they had fevers.

A related but somewhat different mechanism to detect the emergence of pathogens is to look more closely at individual people, specifically the systems that most often host small invaders: the respiratory and digestive systems. The group advanced the idea of using "snot chips" and "smart toilets," both of which have the potential to provide a detectable signal in which to look for pathogenic infection. (The use of these types of devices might raise concerns about privacy because they are more traceable than environmental monitors are, but in the surveillance and monitoring stage, we don't need to record people's identities; we're more concerned with detecting the mere emergence of pathogens.)

Another important way to detect emerging pathogens is with indirect monitoring, specifically, watching the purchases of pharmaceuticals as an indicator of infectious outbreaks. In 1993 doctors learned about a

Cryptosporidium outbreak in Milwaukee only when a perceptive pharmacist noticed a period of particularly strong sales of Imodium (loperamide).

We should develop better informatics and meta-analyses to detect outbreaks like these as early as possible. These bioinformatics tools would also help to synthesize and interpret the loads of biological information gleaned from environmental monitoring. This effort should aim to add to, draw on, or connect similar projects that have already been launched, such as NEON, a National Science Foundation-funded network of ecological observatories that could help with environmental monitoring for pathogens (www.neoninc.org/about/, accessed 2/2/2006); RSVP, an Internet-enabled system that gathers information from thousands of doctors (www.ca.sandia.gov/chembio/implementation_proj/rsvp/, accessed 2/2/2006); and ESSENCE, a Department of Defense-run system to centralize information on infectious outbreaks (www.geis.fhp.osd.mil/GEIS/SurveillanceActivities/ESSENCE/ESSENCE.asp, accessed 2/2/2006).

Detection and identification: Isolating and identifying the cause of an infectious human illness

Once an illness has been detected, scientists need to isolate the cause to determine how best to fight it. Our group especially encourages work in a few fields to improve our ability to identify pathogens. Purification, concentration, and array-based analysis of nucleic acids, proteins, and potentially other molecules should be used more broadly to find the signatures of pathogens. Improving culture methods would accelerate the discovery of infectious agents. We should develop better methods for diagnosing disease states and infectious agents by the reactions of parts of the immune system, such as the T-helper cells. We could also assay the major histocompatiblity complexes (MHCs) of people's cells—not to learn their MHC genotype but to analyze the mixture of peptides in the complexes and see if particular pathogens are evidenced by distinctive protein signatures.

Initial response: Fast strategies for decreasing the harm done in the first stages of an outbreak

Some of the group's most important work came in the identification of technologies that may provide the scientific basis for improvement of initial responses to emerging diseases—the heart of rapid response. Some of the most promising tools in this stage are bacteriophages, which can be used in a couple of different ways. One recommendation is to keep a permanent and openly accessible library of known bacteriophages. After a new pathogenic bacterium is sequenced and its antigens known, it can be compared against the collection of phages to see which might effectively attack it. As the collection grows, it becomes more likely to include effective counters to new pathogens. Bacteriophages can also be used through phage display to make other antibacterial tools, like antibiotic peptides and antibodies for passive immunotherapy.

Just as a widely accessible library of phages can help respond to new pathogens, a library of rejected drug candidates could do the same. Our group recommends maintaining a catalog of compounds that pass phase 1 of the approval process—they are safe—but they fail in phases 2 or 3 because of a lack of efficacy. But these drugs, ineffective for their original intended goal, may well be invaluable cures against emerging diseases. As new infectious agents are isolated, they could be tested against potential treatments in silico; binding and affinity could suggest efficacy. This approach would demand a substantial organizational change—there is currently no way to administer such a project and reward participating drug makers—but we believe it's possible to develop mechanisms to do so. Both the bacteriophage and recycled-drug libraries would demand new bioinformatics tools to speed the connection of potential cures with new pathogens.

A few more of the group's suggestions for initial responses:

- Use T-helper-cell vaccines to direct the immune system to attack pathogens
- Damp down inflammatory responses to certain infections where the inflammatory response is excessive and harmful
- Locate and contain "superspreaders"

Later response: Follow-on treatments for diseases that have not been contained in the previous step

We hope that the aforementioned strategies will help prevent new diseases from causing widespread illness, but it seems all but inevitable that a pandemic will happen again at some point. So the group developed some techniques for treating diseases that have already begun to spread.

Again, some of the useful tools to develop for this stage are bacteriophages. If the structure of the pathogen is known but there are no existing phages to attack it, researchers should work to evolve or engineer a new one to do the job. Phages could also be used to alter the composition of the native microflora to compete with or otherwise hinder the reproduction, spread, or virulence of the pathogen.

The group was also hopeful about accelerating vaccine development using the pathogen's sequence, following on from similar work being attempted on HIV. By exposing patients to DNA plasmids that express pathogen proteins and an engineered adenovirus, we hope to be able to provoke an effective immune response. In an urgent or dire situation, the testing of a vaccine might be accelerated so it could be deployed faster than the usual lengthy approval process allows.

Organizational, social, political context

In rapid response to emerging disease, the organizational and political factors are as important as the science; good organization is not enough to make rapid response work, but bad organization is enough to make it not work. Our group focused on technical questions in our discussions, as that was our obvious strength, but we felt we would be remiss not to mention the societal context.

One great concern is the lack of commercial viability of some important parts of rapid response to emerging disease. Vaccines, for example, are seen as unprofitable for drug companies, and are unfortunately ignored. Government and civil society should work together to ensure that such an important health measure is well covered.

Another problem is the increasing secrecy around important reagents and sequences because of intellectual property protections and demands for secrecy from certain governments. Some group members working on vaccines for H5N1 avian flu have run into problems trying to get information on the virus because of restrictions. Intellectual property and other motivations for secrecy must give some ground to safety in the case of potential epidemics or pandemics.

With much of the effort to improve rapid response focusing on accumulating more information more quickly, this search must also be tempered by ethical concerns, especially the privacy demands of laws such as HIPAA (Health Insurance Portability and Accountability Act).

These are the major points raised by the workgroup in our discussions about rapid response to emerging disease. The group feels that each individual step will most likely improve health care by itself. Moreover, this framework can help make sure that our society won't ignore any important links in the chain of rapid response. As one of the group members points out, "Wal-Mart knows that organization is a technology." We hope that it might help scientists and policy makers to guide research and resources in a way that will best protect global health.

Develop an Inexpensive (and Cost-Effective) Diagnostic Test That Could Be Deployed in Countries with Little Scientific Research Infrastructure

WORKING GROUP DESCRIPTION

Background

Tropical parasitic infections such as malaria, leishmaniasis, and trypanosomiasis, are responsible for millions of deaths per year. Most deaths occur in children or young adults in developing countries. Malaria was eliminated in some regions of the world in the 1960s but remains in Africa, Central and South America, and Asia in strains resistant to chloroquine and other inexpensive and commonly available drugs. Mosquitoes are also developing resistance to common insecticides (Mabey et al., 2004; www.gatesfoundation.org/GlobalHealth/Pri_Diseases/Malaria/default.htm, accessed 2/2/2006).

Point-of-care (POC) diagnostic tests exist for infectious diseases such as malaria. They use immunochromatography to detect antigens or antibodies in a dipstick or lateral-flow format. Companies are manufacturing rapid diagnostic tests for malaria but most have not been carefully evaluated and the performance typically falls below the expected level. The accepted standard for malaria diagnosis remains the evaluation of Giemsa-stained blood smears by light microscopy, which is labor intensive, slow, and requires trained personnel. More widespread availability of simple and accurate dipstick tests for malaria would alleviate a great burden of disease in Africa (Mabey et al., 2004).

The Problem

Diagnostic tests in developing countries need to be widely accessible, inexpensive, and simple to use.

- Determine the constraints necessary to make diagnostic tests widely available, considering such factors as length of test, follow-up visits, supply of reagents, electricity, equipment required, trained technologists, cost, and so on.
- What sensitivities and specificities are required in diagnostic tests to serve large populations?
- Can you take advantage of genomic methods to identify different strains of the disease and use that information to diagnose and ultimately treat individuals on a case-by-case basis and in a cost-effective manner?
- Design a diagnostic device that combines all the desired features you identify in earlier steps. Provide a scenario of how this device can be deployed and used effectively in developing countries.
- How can nanotechnology and new rapid diagnostic methods for other targets be adapted to diagnose malaria species, drug-resistant mutations, and vaccine-resistant polymorphisms in malaria-endemic countries?

If desired, the working group can expand the topic to tropical diseases other than malaria—visceral leishmaniasis, African trypanosomiasis, Chagas' disease (South American trypanosomiasis), and so on.

Initial References

Bashir, R. 2004. BioMEMS: state-of-the-art in detection, opportunities and prospects. Advanced Drug Delivery Reviews 56(11):1565-1586.

Fortina, P., L. J. Kricka, S. Surrey, and P. Grodzinski. 2005. Nanobiotechnology: the promise and reality of new approaches to molecular recognition. Trends in Biotechnology 23(4):168-173.

Gascoyne, P., J. Satayavivad, and M. Ruchirawat. 2004. Microfluidic approaches to malaria detection. Acta Tropica 89(3):357-369.

Jain, K. K. 2005. Nanotechnology in clinical laboratory diagnostics. Clinica Chimica Acta 358(1-2):37-54.

Mabey, D., R. W. Peeling, A. Ustianowski, and M. D. Perkins. 2004. Diagnostics for the developing world. Nature Review Microbiology 2(3):231-240.

Murray, C. K., D. Bell, R. A. Gasser, and C. Wongsrichanalai. 2003. Rapid diagnostic testing for malaria. Tropical Medicine and International Health 8(10):876-883.

Perandin, F., N. Manca, A. Calderaro, G. Piccolo, L. Galati, L. Ricci, M. C. Medici, M. C. Arcangeletti, G. Snounou, G. Dettori, and C. Chezzi. 2004. Development of a real-time PCR assay for detection of *Plasmodium falciparum*, *Plasmodium vivax*, and *Plasmodium ovale* for routine clinical diagnosis. Journal of Clinical Microbiology 42(3):1214-1219.

Singh, N., A. K. Mishra, M. M. Shukla, S. K. Chand, and P. K. Bharti. 2005. Diagnostic and prognostic utility of an inexpensive rapid on site malaria diagnostic test (ParaHIT f) among ethnic tribal populations in areas of high, low and no transmission in central India. BMC Infectious Diseases 5(1):50.

WORKING GROUP SUMMARY

Summary written by:

Susanne McDowell, Graduate Student, Science Communication, University of California, Santa Cruz

Working group members:

- Asem Alkhateeb, Postdoctoral Scholar, Human Genetics, University of Chicago
- Austin Demby, Senior Staff Fellow, Global AIDS Program, Centers for Disease Control and Prevention
- Jeffrey Feder, Associate Professor, Biological Sciences, University of Notre Dame
- Paul Laibinis, Professor, Chemical Engineering, Vanderbilt University
- Colleen McBride, Chief, Social and Behavioral Research Branch, National Human Genome Research Institute
- Susanne McDowell, Graduate Student, Science Communication, University of California, Santa Cruz
- Arcady Mushegian, Director of Bioinformatics, Stowers Institute for Medical Research
- Mihri Ozkan, Assistant Professor, Electrical Engineering, University of California, Riverside
- David Roessner, Senior Evaluation Consultant, National Academies Keck *Futures Initiative*
- Debra Schwinn, James B. Duke Professor, Anesthesiology and Pharmacology, Duke University Medical Center

- Beatrice Seguin, Postdoctoral Fellow, Canadian Program on Genomics and Global Health, University of Toronto
- Diane Sullenberger, Executive Editor, Proceedings of the National Academy of Sciences
- Steven Wolinsky, Samuel J. Sackett Professor of Medicine, Northwestern University Feinberg School of Medicine

Summary

Imagine having to travel dozens of miles from your village or town to reach a health clinic when you get sick. When you finally arrive, you have to stand in line until the doctor is ready to see you. You might then spend another few hours awaiting your test results. Depending on the number of other patients there, you may have to stay the night. And because each clinic provides a limited selection of tests, you could end up repeating this process over several days. This scenario is real for people in parts of Africa and other countries where medical resources are sometimes scarce but where infectious diseases like malaria, tuberculosis, and HIV are rampant. According to the Centers for Disease Control and Prevention, 39.4 million people—mainly in Africa and Asia—were living with HIV at the end of 2004. Tuberculosis, another infectious disease common in developing countries, racked up 2.4 million cases last year. Malaria affected between 350 and 500 million people, 70 percent of whom live in Africa.

With these scenarios and statistics in mind, our working group approached our task: to design a portable, inexpensive test for malaria or other infectious diseases that is easy to administer and that will deliver fast results—within approximately 30 minutes. The test should take into account different disease strains and examine possible drug resistance.

Our group was a motley crew: among them a few geneticists, a chemical engineer, an AIDS expert, a social and behavioral scientist, an infectious disease guru, an electrical engineer, a science journal editor, an evolutionary biologist, and a pharmacology expert. Individually, no one was equipped to devise a diagnostic test with all the properties we wanted. But together, this diverse group of imaginative scientists took only eight intense hours to develop a product that is now in the beginning stages of the patenting process.

Of course, creating this product involved a winding road of brainstorming, questioning, and debate. The major issues on day 1: What disease should we test for (we were charged with malaria but had the option of choosing something else)? How should we tackle sample preparation, assuming those using our device would have limited access to electricity, refrigeration, reagents, and additional equipment? Should we create a protein- or DNA-based test? Could the sensitivity and specificity of our device surpass that of current tests? And how would we incorporate nanotechnology?

We spent a large chunk of our first meeting—and even part of the second—deciding which disease we should target. Initially, we hoped to combine HIV, malaria, and tuberculosis tests on a single platform. But inexpensive, portable HIV diagnostics already exist, and tuberculosis tests involve messy sputum samples requiring more preparation than blood samples.

We decided to focus on malaria, because it would require simple blood samples and would allow us to examine all the challenges outlined in our task. Furthermore, an improved malaria test could fulfill a pressing public health need on a large scale. Though malaria disease complex is transmitted by four main malarial plasmodium species, the fastest and cheapest test on the market does not indicate which of those species is present. Microscopy tests do, but require trained professionals. Furthermore, people who take antimalarial drugs on a frequent basis, as is often the case, harbor pathogens that often develop drug resistance. If we could take these factors into account, we would simplify and accelerate the malaria-testing process and lay the foundation for tests of other infectious diseases.

In the end, we constructed a device composed of two parts: (1) a reusable sample platform that includes a battery, display, and a sample docking port and (2) a disposable sample chip on the order of a few square centimeters that inserts into the platform and performs all the testing necessary for analysis using nanotechnology. Each chip is a miniature laboratory that processes and analyzes the samples.

Developing the platform itself was straightforward, but designing the sample chip necessitated all the expertise in the room. Our hope was that we could load a blood sample onto the chip and initiate a series of reactions that would indicate whether the patient has malaria, the species of plasmodium, and whether the malarial strain is resistant to a particular drug. If necessary, we could create two chips: one that would give a "yes or no" diagnosis, and another that would subsequently provide the details for those diagnosed with the disease.

Among the details that challenged us:

1. *Should we employ a protein- or DNA-based test?* We could test for either antiplasmodium antibodies or the DNA of the malaria parasite. The problem with antibody tests is that they don't distinguish between current or past infection. A DNA-based test can: enough organisms are present to diagnose an active infection. Additionally, DNA tests involve simpler sample preparation methods and diagnostic probes.

2. *How large a sample would we need to get accurate results?* Ideally, the patient would provide a blood sample from a simple finger prick. This sample would adhere to a loop—a ring-shaped device—that would fit securely into the chip. But we debated about how much sample we would need to recognize the parasite. Assuming there are eight to nine plasmodial cells per microliter of blood, 100 microliters would yield about 800 to 900 cells—enough for detection.

3. *How do we separate the DNA from other components in the blood?* We wanted to target the parasite's DNA, so we needed to separate it from other components in the blood, like proteins. To do this we incorporated a filter on the chip itself that would isolate the DNA. Nanotechnology would allow us to pinpoint just the parasitic DNA. We decided to use nanowires—tiny wires smaller than the width of a dust mite. These wires could be equipped with "docks" that attract and anneal only to the parasite DNA. Because nanowires are extremely sensitive, they can detect malarial DNA even if the disease is in the beginning stages and only a few parasites are present.

The result is a "lab on a chip" that would target and examine DNA of malaria parasites. The general procedure is as follows: a blood sample is inserted into the chip, where it rests on top of a small filter. The technician presses a button on the display console to initiate sample processing. A small amount of eluate runs over the top of the filter and carries off the proteins and other waste. A simple vacuum device routes the waste to a miniature "trash bin" on another section of the chip. Another solution unhooks the DNA from the filter and allows it to pass through. Shear forces tear the DNA apart to release the nucleic acids, which are hooked to magnetic tags that route the nucleic acids to the part of the chip outfitted with nanowires. Different nanowires are equipped with different "docking DNA," corresponding with various malarial strains and drug resistance codes. If parasitic DNA is present, it hooks to the matching nanowire. The outcome of the test is displayed on the device console.

The main benefits of this device over other available tests are that it provides detailed results in a short amount of time. We estimate the process will take about 30 minutes—meaning at least 16 people could be tested in a single day, as opposed to four (assuming that a microscope-based test takes at least two hours). Furthermore, patients infected with drug-resistant strains will avoid paying for medicine that will not work and that may exacerbate the drug-resistance problem. And because the testing is confined to a disposable chip, it is safe and easy to perform.

After designing the device, we discussed several unknowns:

- How well the device will actually work
- Whether it is too similar to other new devices (some of which were already being developed independently by members of our group)
- The actual cost of each chip (the group estimated about $5 per chip and $2,000 for the machine and portal)
- How much user training will be required
- How the chips will be disposed of
- How the device itself will be cleaned and how easily it might get contaminated
- Who will provide medical counseling postdiagnosis
- How the device will be distributed

Group members are currently addressing these questions as they seek project funding and apply for a patent.

By the end of the conference we were imagining a new scenario: a technician outfitted with one or two of these diagnostic devices, simple sample collection equipment, and some sample chips visits a village. After taking a small blood sample from the patient, the technician loads the sample on the chip, presses a button, and an entire nanolab gets to work. Less than an hour later the device provides a detailed diagnosis—even if the patient is in the early stages of the disease and has not started to show symptoms.

Our hope is that this malaria test is just the beginning. If it works, we could design other chips that fit into the same platform to diagnose diseases like tuberculosis and HIV. An entire battery of tests could be performed in an office or in the field. Ultimately, we hope the test will save patients and healthcare providers money and time. Above all, we hope it will save lives.

How Would You Spend $100 Million Over the Next Five Years to Prevent the Next Pandemic Flu?

WORKING GROUP DESCRIPTION

Background

The 1918 pandemic influenza killed an estimated 30 million people worldwide. More than 80 percent of the deaths in the U.S. armed forces during World War I were due to the flu rather than combat. Subsequent epidemics have occurred on a regular basis as new flu strains have arisen. Experts agree that without further protective measures, it is only a matter of time before a new and deadly pandemic occurs.

There are currently multiple influenza strains for which humans lack immunity that are circulating in wild bird populations. One of the most dangerous is the H5N1 strain, where H5 and N1 denote the particular variants of the viron surface proteins hemagglutinin and neuraminidase, respectively, which our immune system uses to recognize the virus. Primarily through contact with birds, the H5N1 strain has infected more than a hundred people with a high mortality rate but has not yet gained the capacity for efficient transmission from human to human. In previous epidemics such capacity has often been achieved by a process called viral reassortment, in which some of the eight segments of the influenza genome are exchanged in the cells of a pig that is simultaneously infected with a deadly avian strain and a common mammalian strain. Population growth in Asia has greatly increased the number of locations where domesticated birds, pigs, and humans are living in close quarters, providing more opportunities for reassortment than before, and increased worldwide travel has provided better opportunities for a new reassorted flu strain to develop into a pandemic. Hence, the gloomy prognosis by the experts.

The Problem

The challenge to the working group is to come up with a strategy to prevent the next flu pandemic, be it H5N1 or another newer strain. Some possible approaches are:

1. Development and deployment of antivirals, such as oseltamivir phosphate (Tamiflu), which blocks the essential action of neuraminidase, inhibiting the mature viron particles from exiting the infected cell, at least until a resistant flu strain appears
2. Development of rapid means of creating and deploying a new vaccine that is specific to a new flu strain
3. Building an early warning system that is capable of detecting a new potential pandemic flu strain before it has infected so many people that containment is difficult
4. Develop a better understanding of the human host immune system to allow completely new types of intervention

Working group members are encouraged to explore ways genomics can help in these tasks. When considering detection, the group might consider a technology spectrum from simple antibody tests to full sequencing of flu genomes, exploring which systems could be effectively deployed in the field in the third world and which procedures would be centralized. Use of sophisticated research tools, such as reverse genetics, should be considered. Here *E. coli* plasmids containing the eight segments of the flu virus plus some viral proteins to initiate viral replication are transfected into mammalian cell lines so that different virus strains can be created and propagated in cell culture. This method is increasingly used to create live attenuated vaccines. Working group members should also contemplate the dangers of misuse of these technologies.

Initial References

Baulch, V. When the flu ravaged the world. Detroit News, Online: at info.detnews.com/history/story/index.cfm?id=116&category=events, accessed 2/2/2006.

Fauci, A. S. 2005. Race against time. Nature 425(May 26):423-424.

Fouchier, R., T. Kuiken, G. Rimmelzwaan, and A. Osterhaus. 2005. Global task force for influenza. Nature 425(May 26):419-420.

Ho, D. 2005. Is China prepared for microbial threats? Nature 425(May 26):421-422.

Normile, D. 2005. Who controls the samples? Science 309(Jul. 15):372-373.

Osterholm, M. T. 2005. A weapon the world needs. Nature 425(May 26):417-418.

Webster, R., and D. Hulse. 2005. Controlling avian flu at the source. Nature 425(May 26):415-416.

WORKING GROUP SUMMARY

Summary written by:

Haley Poland, Graduate Student, Annenberg School of Journalism, University of Southern California

Working group members:

- Myles Axton, Editor, Nature Genetics
- Robert Carlson, Senior Scientist, Electrical Engineering, University of Washington
- Roxanne Ford, Program Director, W. M. Keck Foundation
- David Haussler, Director of the Center for Biomolecular Science and Engineering and Professor of Biomolecular Engineering, Howard Hughes Medical Institute, University of California, Santa Cruz
- Stephen Albert Johnston, Director and Professor, Center for Innovations in Medicine, Biodesign Institute at Arizona State University
- Kam Leong, Professor, Biomedical Engineering, Johns Hopkins University
- Haley Poland, Graduate Student, Annenberg School of Journalism, University of Southern California
- Karin Remington, Vice-President, Bioinformatics Research, J. Craig Venter Institute
- Christina Smolke, Assistant Professor, Chemical Engineering, California Institute of Technology
- Lyna Zhang, National Center for Infectious Disease, Centers for Disease Control and Prevention

Summary

Working group members had their work cut out for them. With $100 million in imaginary (virtual) money and a dream, they set out to prevent the next pandemic influenza. In the past several years H5N1, an influenza virus endemic to wild bird populations, has infected more than 100 humans. While the virus has not yet mutated to become transmissible between humans, its potential to do so has scientists scrambling to derive new methods and approaches. This group of scientists had quite a plan.

Simply stated, $100 million is probably not enough money to prevent a global pandemic. Though advised by conference administrators that more virtual money was available if it would make the proposed challenge more feasible, the group decided that limited funds would encourage a focused and prioritized plan of action. As the group members brainstormed to organize the spending in a way that would complement currently proposed flu preparedness plans, the far-reaching impact of limited but well-placed funding would become clear.

Following eight hours of intense collaboration through roundtable discussions, the money was divided into two broad categories. Approximately half would fund a centralized flu research resource. The remaining half would finance a series of grants and contracts for vaccine and antiviral research, excluding human trials and full-scale vaccine production. Group member Rob Carlson explained how the action plan could be sustained once the $100 million was exhausted. "This money will make the initial stages go faster," he said, "and then when things work, it will be obvious what should garner more funding."

Centralized research and resource facility

Steered unobtrusively by the group's administrator, David Haussler, the group decided the primary goal should be to promote innovative, genome-centered research. The group members agreed that a core research facility dedicated to gathering all available flu information in a centralized location was imperative for rapid and cost-effective vaccine development.

The planned community flu-research resource would run on a $5 million administrative budget. It would dedicate $5 million to sequencing full genomes of the bird and human versions of the virus, as well as to monitoring emerging strains. With $15 million, the plan would also provide access to BSL3+ and BSL4 testing facilities for a broader group of scientists to test for the virulence and transmissibility of threatening mutation combinations, and the effectiveness of new vaccines and antivirals. An international clone library would be established to consolidate vaccine and antiviral research, as well as provide plasmids coding for known gene variants to vaccine development laboratories, discussed below.

If a virus like pandemic flu surfaces, global spread is essentially unavoidable because most people have little to no immunity. Because it would spread like wildfire, detection just one day earlier could have dramatic effects. Monitoring people for viral antibodies, which show up a few days after infection, is often too late. "Antibodies are historical information," quipped Stephen Johnston. "They'll tell you what someone died of."

The group therefore spent another $5 million, distributed across five laboratories, on the development of new field detection technologies. Ronald Davis, director of the Stanford Genome Technology Center, briefly joined the group to discuss inexpensive and rapid detection devices in the works. He said that while the technology to identify specific flu strains is there, such as with the recently developed flu chip, the readout devices present the greatest obstacle to fast, on-site detection. He seemed certain, however, that increased funding would make the difference. "It's not like building a bomber," he said. "One million dollars can do so much."

Changes in evolutionary genetic and ecological factors of a virus are inextricably linked to its virulence and pathogenicity. Consequently, the group allocated $10 million to modeling the landscape of flu strains edging toward pandemic by documenting how the virus is changing over time in its wild reservoirs, and how it might mutate or combine with other viruses in the future according to observed patterns.

Myles Axton presented an analogy comparing the evolutionary landscape of a flu virus to a golf course. With the hole signifying a devastating pandemic flu, the green surrounding the hole represents all the versions of that virus that an ideal vaccine should protect against. The ridges and slopes of the golf course are the characteristics of the virus's environment that determine whether it rolls toward or away from pandemic potential. Cataloging and understanding these bioinformatics (ridges and slopes), which would be a key purpose of the community flu resource, is necessary in the creation of an anticipatory and broadly protective vaccine. "We have to think evolutionarily," said Axton. "Otherwise we're just playing catch-up with a virus that can clearly outrun us."

The Gimish *vaccine competition*

Central to the group's plan was the *"Gimish"* vaccine concept: create a vaccine that protects against many known variants of the virus, as well as anticipates possible mutations.

Traditional vaccines consist of attenuated (weakened) or killed viruses that stimulate the immune system's humoral response. The antigens introduced by the vaccine "instruct" B cells to produce antibodies with the help of T cells; the antibodies adhere to the antigens and flag them for destruction by white blood cells. The immune system is then "trained" to deal with the active virus in the future. Annual influenza vaccines are typically killed viruses grown in eggs or cell culture. Because a pandemic flu is both yet unknown and rapidly evolving, an attenuated or killed virus vaccine is insufficient.

Another method of influenza vaccine development involves reverse genetics, a process by which cloned DNA is custom arranged to code for only certain flu antigens that will trigger immune response. Synthetic vaccines range from these expressed protein subunit vaccines to full gene vaccines in which "naked" DNA plasmids with genes coding for pathogenic proteins are directly injected.

Gene vaccines differ from traditional vaccines in that they trigger both humoral and cellular immune responses. Gene vaccines work by introducing a gene that codes for an antigenic protein directly into the nucleus of dendritic immune cells. When the gene is expressed, the surface of the cell is modified in such a way that cellular immune response is triggered: white blood cells recognize and kill foreign organisms and infected cells as detected by surface proteins. The humoral response is also triggered because antibodies respond to the flu antigen(s) secreted by those altered cells.

A *Gimish* vaccine for pandemic flu, using the virus landscape model as systematized by the community research center, would incorporate as many potential variants of the virus as possible using gene vaccine technology. In this case, researchers would also engineer site-directed mutations in potentially pandemic flu viruses (also through reverse genetics) and evaluate how virulence and pathogenicity are affected. By genetically engineering what one group member called "our own nasty version of the virus," as informed by the bioinformatics models, the central research center could test all vaccines created by the competing labs and gauge how the flu variants respond.

Fifty million dollars was earmarked to subsidize parallel research efforts to produce a *Gimish* gene vaccine. For the *Gimish* vaccine competition, which one group member called the "elimination jamboree," 10 laboratories would be selected to receive $1 million, one-year grants to work toward a fundamentally new, broadly protective gene vaccine. All labs would have access to the established community research resources, and by interfacing with that infrastructure, could save considerable time and money.

At the end of one year, two standout labs would each be granted $20 million, two-year contracts to create a *Gimish* vaccine. The two chosen labs would have to demonstrate the capacity and capability to meet the following criteria: 100 million vaccine doses costing less than a dollar each produced in two months; broad protection for two animal species; suitable for distributed global production; and broad protection against many virus variants and, ideally, anticipate future changes as well.

The development of adjuvants to optimize a *Gimish* vaccine would also be encouraged. Adjuvants are substances that enhance the immunogenicity of antigens, meaning that when they are administered in combination with another treatment, they improve the body's immune response. In this case, adjuvants could promote broader protection and vaccination in one dose.

In addition to grants for a *Gimish* vaccine, $5 million was allocated for seed grants for antiviral research. One new antiviral approach halts viral proliferation by capitalizing on a natural cellular process called RNA interference (RNAi). When a virus enters a cell and begins replicating, a nuclear enzyme known as Dicer cleaves the viral RNA into short interfering segments (siRNA). These siRNA segments are picked up and unwound by a protein complex, and then essentially re-adhere to viral RNA strands and prevent their replication. Antiviral therapy involves introducing artificial siRNA into potential host cells, essentially tricking the cells into activating the RNAi pathway before the virus is actually present. This is promising for reducing influenza pathogenicity and transmission for two reasons: it requires only a synthesized segment of the viral genome (which could be selected from those already sequenced) and it does not require waiting for the virus to appear in the population. The primary problem with RNAi in flu prevention is its requirement for prepositioning of the siRNA in a cell before the virus infiltrates the body. Because pandemic flu strains like H5N1 do not target specific cells, delivery is highly problematic.

The group decided to award ten $500,000, one-year grants to laboratories committed to ongoing siRNA research, with a deserving laboratory receiving $5 million upon demonstrating a potentially effective RNAi therapy.

Perhaps the most crucial characteristic of the group's comprehensive plan was the feedback loop. Ongoing sequencing of flu strains would not only drive vaccine production but also allow for constant updates of the detection devices. As vaccines from the competing labs were to be tested against the constructed viruses, the efficacy or inefficacy of those vaccines would inform both the landscape model and ongoing research. Similarly, advances in RNAi therapy could shift the research center's focus from the *Gimish* to methods of siRNA introduction. The research findings of each laboratory would become commonly available through the centralized facility, with an understanding that thwarting the next pandemic flu requires greater openness in research while still meeting the biosecurity needs of participating governments.

At the conclusion of eight hours of lively conversation and planning, this brain conglomerate had contributed its fair share. Many of the energized group members seemed genuinely disappointed that they had been dealing with virtual money all along. While the $100 million challenge may have seemed broad and amorphous at the start, it had been whittled into a sharply focused vision. With the momentum of the collaboration pushing them along, several group members refused to let the plan evaporate on the last day. Rumor has it that a grant proposal may be in the works to help bring them together once again.

How Can Genomics Facilitate Vaccine Development?

WORKING GROUP DESCRIPTION

Background

Vaccines are the most efficacious means of minimizing the impact of infectious diseases on the human population. The challenges and importance of making vaccines that will meet FDA approval have never been greater. Genomics has the potential to improve the process of vaccine development substantially. Genome sequencing can help to identify genetic patterns related to the virulence of a disease, as well as genetic factors that contribute to immunity or successful vaccine response. All this information could lead to vaccines with better and more specific targets that elicit more successful protective immune responses. Comparing the genome sequences of viruses that cause infection with those that do not may provide additional insights. In turn, genome manipulation can facilitate derivation of attenuated strains or other vehicles for delivery of the desired antigens to stimulate immune response. On the other end of the spectrum, analysis of host diversity can reveal effective immune responses and possibly the genetic basis for inappropriate response. The recent progress in definition of the innate immune system, necessary for acquired response, should facilitate the definition of this host diversity.

The Problem

Explore the ways these and future approaches in genomics might be applied to speed the development of vaccines. Targets include emerging threats from either natural reservoirs or terrorist activities; established targets for which there are either no or only ineffective vaccines; and pathogens with effective vaccines with unacceptably high rates of untoward reactions. Beyond genome sequence analysis, participants should consider related technologies, such as the analysis of gene expression by either the pathogen or host upon infection or vaccination; proteomic analysis, including protein-protein interactions within the pathogen or between host and pathogen; pathogen and host rapid phenotyping, whole genome synthesis; and the design of more effective vaccine vehicles and adjuvants.

Initial References

Fauci, A. S., N. A. Touchette, and G. K. Folkers. 2005. Emerging infectious diseases: a 10-year perspective from the National Institute of Allergy and Infectious Diseases. Emerging Infectious Diseases 11(4):519-525.

Rappuoli, R. 2004. From Pasteur to genomics: progress and challenges in infectious diseases. Nature Medicine 10(11):1177-1185.

Relman, D., and S. Falkow. 2001. The meaning and impact of the human genome sequence for microbiology. Trends in Microbiology 9(5):206-208.

Scarselli, M., M. M. Giuliani, J. Adu-Bobie, M. Pizza, and R. Rappuoli. 2005. The impact of genomics on vaccine design. Trends in Biotechnology 23(2):84-91.

Segal, S., and A.V. Hill. 2003. Genetic susceptibility to infectious disease. Trends in Microbiology 11(9):445-448.

WORKING GROUP SUMMARY

Summary written by:

Cecilia Dobbs, Graduate Student, Science Journalism, New York University

Focus group members:

- Dat Dao, Director, Life Sciences and Health Group, Houston Advanced Research Center
- Cecilia Dobbs, Graduate Student, Science Journalism, New York University

- Ananda Goldrath, Assistant Professor, Biology, University of California, San Diego
- Muin J. Khoury, Director, Office of Genomics and Disease Prevention, Centers for Disease Control and Prevention
- Corinne Lengsfeld, Associate Professor, Engineering, University of Denver
- Alan McBride, Researcher, Gonçalo Moniz Research Center, Oswaldo Cruz Foundation, Salvador, Brazil
- Catherine McCarty, Interim Director and Senior Research Scientist, Center for Human Genetics, Marshfield Clinic Research Foundation
- Fabienne Paumet, Associate Research Scientist, Physiology and Biophysics, Columbia University
- John E. Wiktorowicz, Associate Professor, Human Biological Chemistry and Genetics, The University of Texas Medical Branch
- Zhenhua Yang, Assistant Professor, Epidemiology, University of Michigan, School of Public Health

Summary

If the developing world could run classified ads, one ad would certainly read, "Needed: vaccines, fast and cheap." In 2005 alone, if there had been a vaccine for malaria, it could have saved more than a million people's lives. If there had been a vaccine for HIV, it could have protected more than 3 million people from infection. If there had been a method of manufacturing influenza vaccines that did not need 12 months and hundreds of millions of chicken eggs, many of the current fears over pandemic flu could be allayed. Each case is different, but the bottom line remains the same: if we are ever to triumph over infectious disease, vaccine development needs to become faster and cheaper.

It takes between 7 and 15 years and $200 million to $600 million to develop a vaccine. In many cases, that financial investment is well worth it. Vaccines still represent one of the most cost-effective medical interventions for preventing death from disease, and hundreds of millions of lives have been saved over the decades because of them. Yet the basic approach to development has changed very little during that time, and scientists have had little or no success in combating diseases like malaria, TB, and HIV. Clearly, new approaches are needed for old problems.

One promising source for innovation is the burgeoning field of genomics and proteomics. Current and future technologies in these fields will enable scientists to study an organism's entire set of genes and proteins simultaneously, instead of working with one gene or protein at a time. This could help identify genes or proteins that play a key role in a pathogen's ability to infect and the host's immune system response, and ultimately lead the way to better vaccines. Exactly how much could the fields of genomics and proteomics enhance the process of vaccine development? At the third annual National Academies Keck *Futures Initiative* a group of nine researchers, with backgrounds ranging from bioengineering to immunology, came together to tackle this very question.

In the case of infectious diseases, of course, there are actually three genomes or proteomes to consider: that of the host, the pathogen, and the vector (like the mosquito in the case of malaria). When it comes to studying the host genome, some people are naturally more resistant to disease than others, and scientists may be able to identify genes or portions of genes that play a role in that immunity. Alternatively, data from the pathogen genome may help scientists identify the genes that play a key role in the pathogen's ability to infect and harm the host. Vaccine development begins with epidemiology and pathogenesis—information gathering about the disease itself. Before researchers can begin to develop a vaccine, they need to know what type of pathogen causes the disease, how it is transmitted, what cells it targets in the body, and what negative effects it has on the host. Currently, most epidemiology efforts, particularly with emerging diseases, focus entirely on the characteristics of the pathogen, and do not take genetic variables in the host into account.

The more data collected, the easier it is to develop an appropriate and properly targeted vaccine. Some of the major stumbling blocks to vaccine development could be overcome if better and more complete information existed on the genetic variability of pathogens, as well as on the genetic basis for host susceptibility or resistance to disease. On the proteomics side, more data could enable scientists to identify cellular responses to infection, as well as biomarkers for infection and for successful vaccine response. The group concluded that vaccine development research needs a better infrastructure than there is right now for data collection that would include information in populations on both pathogen and host, and specifically one that would record the type of immune response, if any, elicited in the host, as well as any genomic or immunological markers that may aid in properly targeting the vaccine.

A second shortcoming in the basic research stage occurs in the laboratory. The pathogen needs to be culturable in the laboratory, but the conditions for growth are often very specific and difficult to determine. Sometimes this means that an organism simply will not grow in culture, but in other instances the culture conditions actually create selection pressures that result in mutated strains of the pathogen that do not exist in nature and are therefore useless for developing a vaccine.

A similar challenge exists for scientists who are trying to crystallize proteins to study their shape, because the conditions for crystallization vary widely with different types of proteins and are also difficult to achieve. A member of the group suggested that a technology used in protein crystallization could be applied to culturing pathogens, and the concept of the nanoarray incubator was born. This incubator would have several intersecting reservoirs, each inputting a different component of the growth environment, thus creating multiple compartments, each with a unique growth environment. Thus, a variety of growth environments could be tested simultaneously and the correct conditions could be identified more quickly.

Once an organism has been successfully cultured in the laboratory, the next step in developing a vaccine is identifying which of the pathogen's proteins should be the target antigens. This is a monumental task; there are many, many possibilities, and there is no easy way to determine which antigens will elicit the right type of immune response or a strong enough immune response. This involves a lot of trial and error, but the group determined several possible ways that genomics might be able to improve both the speed and the accuracy of the process.

High-throughput methods, in which multiple antigens can be screened simultaneously, could dramatically speed the process. In this way, a scientist can study and characterize many antigens in the same amount of time that it previously would have taken to study just one antigen. An existing process, called reverse vaccinology, begins with the sequenced genome of a pathogen, and then uses statistical analysis to identify the genes that are most likely to influence the pathogen's ability to infect the host. The proteins that these genes code for become the target antigens, and a vaccine is created from this information. None of the vaccines currently available were developed this way, but the group felt that using genomic information this way has the potential to dramatically streamline the vaccine development process in the future. A similar approach could be used with proteomics: sequence the entire proteome of the pathogen and then target that whole protein combination, instead of just a single antigen or two.

After selecting the target antigens, the next step in the process is testing the vaccine in animal models to see whether it is both safe and effective. Unfortunately, animal models are far from perfect. It is possible, even likely, that a good result in animals will not translate into a good result in humans. The group suggested that with genomic analysis, future animal trials may be able to use biomarkers to quickly identify whether a vaccine is effective or toxic, or maybe even identify genetic markers that predict how effective a vaccine will be in humans. Other markers might be able to identify whether the vaccine will convey protective immunity.

Exploring even further outside the box, genomic advances may someday make animal models obsolete. If scientists can use genomic data to engineer human tissue in the laboratory, they may be able to study host-pathogen interactions and vaccine efficacy directly in human tissue and eliminate the guesswork of translating results from animal to human models.

After all the epidemiological studies, searches for antigens, and animal modeling, a vaccine is finally ready for clinical trials. This is by far the most expensive part of the process, and often trials go on for some time before discovering that the vaccine is ineffective or has unforeseen side effects. Here, too, the group felt genomics could improve the process. It may be possible to identify and test for biomarkers that alert researchers to toxicity or efficacy issues early in the clinical trial process. It may also be possible to identify the most genetically susceptible populations for a particular disease and thereby reduce the sample size needed for an effective clinical trial, which would also reduce the cost.

If clinical trials determine that a vaccine is both safe and effective, it is approved and released to the market. But problems often arise in the real world that did not show up in the controlled experiment. A vaccine may work only for one particular population group, or may be harmful for some other population group. The result is usually that the baby is "thrown out with the bathwater" and a vaccine that could have been very beneficial to some people is discarded entirely because of negative effects in other people. This happened in 1999 when a vaccine for rotavirus was pulled from the market after a small number of children who received the vaccine developed intussusception, a form of bowel obstruction. The vast majority of children who received the vaccine never developed this intestinal problem, so the vaccine was clearly good for some people but not for others.

Here again, the group felt genomic information could help. There may be genetic markers that could identify children for whom the vaccine would be effective and those for whom it would likely cause problems. But very little data is gathered once a vaccine hits the market, meaning that both successes and failures in vaccine development do not inform future work. Just as more data collection is needed at the beginning of the vaccine development process, more information is also needed at the end of the process (postmarket surveillance).

There are several existing databases that could be expanded and used to gather important data on vaccines. For instance, the Vaccine Safety Datalink is a program run by a network of HMOs in conjunction with the National Immunization Program at the Centers for Disease Control and Prevention. It already collects data on about 5 percent of the U.S. population, and could be expanded to collect genomic information regarding immune response to a vaccine. If a larger percentage of the population were registered with this database, researchers could track the effectiveness of vaccines in the real world after they hit the market. The information gleaned from this monitoring could feed back into the start of the process, informing and improving the development of future vaccines.

At nearly every stage of vaccine development, the group was able to identify a number of ways that genomic information could accelerate the production of good vaccines while keeping costs low. Scientists have been searching for vaccines for malaria, tuberculosis, and HIV for many years now, and the classic approaches have yielded few results. But the group was confident that with new information from genomics and proteomics, better systems of development are on the horizon, and we hope new, successful vaccines are as well.

Develop a Device to Rapidly and Sensitively Detect and Identify Pathogens in an Environment or Population, Spread Either Naturally or Through Deliberate Acts

WORKING GROUP DESCRIPTION

Background

If not detected and treated promptly, numerous emerging and reemerging infectious diseases (such as cholera, dengue fever, malaria, SARS, and West Nile virus), biological weapons (such as anthrax, smallpox, botulism, and bubonic and pneumonic plagues), and chemical weapons (such as ricin, sarin, and cyanide) have the potential to cause devastating public health crises that could result in the loss of millions of lives. Global travel by millions of people each year accelerates spread of disease, making it even more critical that new rapid-detection methods are devised and validated. To address this threat there is a need for rapid assay strategies for use in clinical diagnostics and environmental detection.

Conventional methods for identifying biological agents (such as immunologic assay and culture) and chemical agents (such as mass spectrometry) generally require high concentrations of the agent, involve complex labor-intensive processing, utilize several pieces of laboratory equipment, and must be executed by trained laboratory personnel. Such genomic methods as quantitative polymerase chain reaction (PCR) and microarrays improve the speed and sensitivity of diagnosis and increase the number of assayed markers (resolution) but still require trained personnel and are expensive.

The Department of Homeland Security and the Department of Defense are funding the development of environmental detectors that monitor outdoor air for biologic and chemical weapons. Sampling of air particles 1 to 10 microns in size is performed by vacuum, centrifuge, or tiny jets, whereas isolating and identifying bacterial, viral, or toxic particles use immunoassays, PCR, or mass spectrometry screens. These approaches have trade-offs that include speed, sensitivity, and cost (Brown, 2004). Fully automated systems capable of unattended collection, sensing, analysis, and reporting remain elusive.

The Problem

- Is it possible to build a device to rapidly and sensitively detect and identify such pathogens as bacteria, viruses, or toxins in an environment or population spread either naturally or through deliberate acts?
- Can genomics help differentiate between natural and deliberate disease outbreak and provide evidence for attribution?
- Can sensors quickly, cheaply, and accurately detect one of the dozens of bacteria, viruses, or toxins that could become aerosolized bioweapons? (Brown, 2004)
- Is it more effective to perform syndromic surveillance of patients in hospital emergency rooms?
- What is the role of such factors as cost, sensitivity, speed, complexity, minimum time needed to detect and identify a pathogen, and false positive rates, and what should be done with the information?

Initial References

Andreotti, P. E., G. V. Ludwig, A. H. Peruski, J. J. Tuite, S. S. Morse, and L. F. Peruski Jr. 2003. Immunoassay of infectious agents. BioTechniques 35(4):850-859.

Bashir, R. 2004. BioMEMS: state-of-the-art in detection, opportunities and prospects. Advanced Drug Delivery Reviews 56(11):1565-1586.

Brown, K. 2004. Up in the air. Science 305:1228-1229.

Fortina, P., L. J. Kricka, S. Surrey, and P. Grodzinski. 2005. Nanobiotechnology: the promise and reality of new approaches to molecular recognition. Trends in Biotechnology 23(4):168-173.

Hahn, M. A., J. S. Tabb, and T. D. Krauss. 2005. Detection of single bacterial pathogens with semiconductor quantum dots. Analytical Chemistry 77(15):4861-4869.

Ivnitski, D., D. J. O'Neil, A. Gattuso, R. Schlicht, M. Calidonna, and R. Fisher. 2003. Nucleic acid approaches for detection and identification of biological warfare and infectious disease agents. BioTechniques 35(4):862-869.

Peruski, L. F. Jr., and A. H. Peruski. 2003. Rapid diagnostic assays in the genomic biology era: detection and identification of infectious disease and biological weapon agents. Biotechniques 35(4):840-846.

Slezak, T., T. Kuczmarski, L. Ott, C. Torres, D. Medeiros, J. Smith, B. Truitt, N. Mulakken, M. Lam, E. Vitalis, A. Zemla, C. E. Zhou, and S. Gardner. 2003. Comparative genomics tools applied to bioterrorism defense. Briefings in Bioinformatics 4(2):133-149.

WORKING GROUP SUMMARY

Summary written by:

Jonathan Stroud, Graduate Student, Annenberg School of Journalism, University of Southern California

Working group members:

- Mary Jane Cunningham, Associate Director, Life Sciences and Health, Houston Advanced Research Center
- George Dimopoulos, Assistant Professor, Molecular Microbiology and Immunology, Johns Hopkins School of Public Health
- Robin Liu, Manager, Microfluidics Biochip, Combimatrix Corporation
- Dan Luo, Assistant Professor, Biological and Environmental Engineering, Cornell University
- Deirdre Meldrum, Director of the NIH Center of Excellence in Genomic Science (CEGS) Microscale Life Sciences Center and Professor of Electrical Engineering, University of Washington
- George O'Toole, Associate Professor, Microbiology and Immunology, Dartmouth Medical School
- Jonathan Stroud, Graduate Student, Annenberg School of Journalism, University of Southern California
- William Sullivan, Professor, Molecular, Cell and Developmental Biology, University of California, Santa Cruz
- Joseph Vockley, Laboratory Director, Life Sciences Division, Science Applications International Corporation
- Debra Weiner, Attending Physician, Emergency Medicine, Children's Hospital Boston, Assistant Professor of Pediatrics, Harvard Medical School

• Lloyd Whitman, Head, Code 6177, The Surface Nanoscience and Sensor Technology Section, Naval Research Laboratory

• John Wikswo, Gordon A. Cain University Professor, Vanderbilt Institute for Integrated Biosystems Research and Education, Vanderbilt University

Summary

There are six questions to ask: Who, What, Where, When, Why, and How? Five out of six can be easily answered:

The what: detecting and identifying the biologic agents and toxins that cause disease
The where: anywhere there is the potential for disease
The when: as soon as possible
The who: the general public
The why: the easiest of all—to save lives

But how? That was the question that a 10-member working group attempted to answer at the third annual National Academies Keck *Futures Initiatives* Conference in Irvine, California. The group, consisting of doctors, scientists, executives, and engineers, worked to outline an approach to find the pathogens responsible for disease and effectively identify them, a technological challenge that if successful, could help prevent pandemics and minimize the effectiveness of future bioterrorist attacks.

Genomics, the science of deciphering or reading the genetic alphabet, was the focus of the conference, and so the group utilized this rapidly advancing field, building strategies and methods to detect and identify biologic and toxic agents capable of producing disease.

"Genetics isn't the only thing we need to analyze," Professor John Wikswo said. "But right now, it's the only thing that has the requisite breadth. That's because genetic sequencing is the only science to yield complete information about activity at the cellular level."

Instead of focusing on one specific disease or pathogen class, the group decided to develop a matrix—a decision tree flowchart—for addressing all possible pathogens and diseases in a variety of scenarios. This would also include the capability to detect pathogens in their insect vectors, such as mosquitoes and ticks, stressed Dr. George Dimopoulos, an assistant professor at the Johns Hopkins School of Public Health.

The group laid out the broad gaps in current biomedical technologies, the proposed problem, and the various difficulties associated with broadening the application of genomics in disease detection. What do we need to measure? they asked. What technologies exist that we could use? What still needs to be developed or integrated? What might we consider in the distant future?

The group moved on to delineate systems criteria—a set of characteristics any device or method would need—and began to address the problem. Each member offered his or her opinion of ideal criteria and based on the conference call discussion, developed a matrix—or chart—to divide up and lay out what the group would delve into over the course of the conference.

"Ideally, we want to have something to detect the pathogens in 30 minutes or less," said Dr. Robin Liu, the manager of microfluidics biochips for the Combimatrix Corporation. Discussion ensued on the basic requirements, with each member in agreement on basic parameters.

"Ideally, we'd like a device that's highly specific and sensitive, can be multiplexed to detect multiple markers, is fully integrated, automated, and portable, has low power consumption, and is disposable," Wikswo reiterated. A modest goal, indeed.

The group decided DNA and RNA (nucleic acid) assays and miniaturized microarrays were the best option at present for analyzing the disease triggers in the body or environment. The problem with DNA-based assays is that most DNA assays require killing cells or organism samples to harvest their genetic material—somehow you have to go from an intact organism to DNA material.

Ultimately, direct nucleic acid assays won out over the more conventional amplification assays that require the PCR to obtain adequate genetic material, or immunoassays that require recognition of specific molecular structures of a particular pathogen or agent.

"You're not easily going to be able to detect a signal pathogen in most samples with an immunoassay," said Lloyd Whitman, a section head at the Naval Research Laboratory.

"I think it's pretty clear that nucleic-acid-based detection will be more effective," agreed Dan Luo, an assistant professor of biological and environmental engineering at Cornell University.

Ultimately, they focused on the three main knowledge gaps identified by Whitman that must be overcome in order to achieve the stated goal: (1) the development of ubiquitous, multivariate sensing, which means an all-encompassing sensor network that comprehensively monitors agents associated with disease; (2) the design of data processing and bioinformatics capabilities to analyze and integrate the information obtained from the sensors; and (3) the improvement of not only our understanding of disease transmission and immune responses but also how these both enable and limit pathogen detection.

Day 2 began with a discussion of whether detection should be clinic- or environment-based. The clinical approach would test people with infections—on the order of thousands of people. The environmental approach would measure agents in the environment, thus targeting entire populations—on the order of millions of people—who were potentially at risk for disease but not necessarily infected. The group evaluated the relative importance, efficiency, and practicality of these two approaches.

The group viewed the problem from the perspective of biodefense, pandemic prevention, and improvements to health care, with particular attention being paid to existing technologies in biodefense and health care. The group ultimately identified a broad approach that would capture all these cases and address the current technology gaps in the process of treating an outbreak, a pandemic, or even the common cold.

To address and organize what any future system might require, they began to devise a decision tree—a basic flowchart—that would describe a comprehensive process for detecting disease-causing agents.

A broad detection strategy must address the possibility that pathogens might be found in any of four main target areas: (1) in the environment (drinking water, air, soil); (2) in people (at the doctor's or in the field); (3) in animal populations that serve as vectors for infectious agents (as with avian flu); and (4) airports, seaports, and border checkpoints that serve as entry portals.

"You find a bird that is sick and must ask whether it has bird flu or another bug, or a patient with a bad cough, an international traveler with a fever, or gray goo on the ground. Are there serious pathogens present?" asked Wikswo.

The group then refined its decision trees to outline how a researcher or physician might proceed when faced with an unknown symptom in people or animals, eventually covering both detection and response. By also analyzing possible procedures for pathogen detection in the environment or in the water or food supplies, they succeeded in identifying two technologies required to span existing technology gaps.

"For each of the four targets, you end up needing two things: a microbe identifier and a pathogenicity analyzer," Wikswo said.

They then set to work discussing and outlining each object and its relevant challenges. First, a microbe identifier would identify genetic material and compare it to a database of known pathogens. Studying the host response to the pathogen includes analyzing changes in host gene expression. Since the most effective (but not necessarily the fastest) way to identify the pathogen would be through its genome, this effort would clearly benefit from the $1,000 genome sequencer under consideration by other conference working groups, a technology the whole group seemed to believe would be available in the near future.

Examples of pathogen identifier technologies considered by the group included Robin Liu's proposed Combimatrix continuous bioaerosol monitor and the Compact Bead Array Sensor System (cBASS) being developed by the Navy.

Second, in order to deal with a previously unidentified pathogen, the group discussed the capabilities required for a pathogenicity analyzer—a machine that would identify the ability of unknown or hacked pathogens to cause disease in cells, model organisms, animals, or people. The group considered a number of technologies, including gene arrays, electrochemical sensing of cellular metabolism, optical or magnetic immunoassays, cellular fluorescence, and flow cytometry—all techniques that could detect pathogen-induced changes in cellular function.

"I'd say you want to look for as many things as you can and then fuse them all together," said Whitman.

Coupled with the required advancements—ease of use, low rates of false positives and negatives, small size, ease of deployment, and relatively low cost—these two devices, it was determined, could revolutionize health care and disease detection and prevention.

"Screening this way could improve diagnosis and treatment of infections as well as prevent infection," said Debra Weiner, attending physician at Children's Hospital Boston and assistant professor at Harvard Medical School.

"Having this kind of approach would help the implementation and development of novel antibiotics that are not based on killing the bacteria," said George O'Toole, associate professor in the Department of Microbiology and Immunology at Dartmouth.

After the final presentation, the group members agreed that their work addressed a very important issue and that the presentation was a great cap to an important set of discussions.

In summary, the conference gave these researchers a chance to share knowledge and ideas and led to crosstalk that in the future may lead to an all-encompassing approach to a difficult problem.

Are There Shared Pathways of Attack That Might Provide New Avenues of Prevention?

WORKING GROUP DESCRIPTION

Background

Microbial pathogens proceed through a series of general steps upon infecting hosts. A pathogen senses the host through surface receptors, which in turn trigger changes in the pathogen. These changes might include remodeling its surface and expressing proteins that adhere to the host or facilitate entry. Pathogens also deliver proteins and small molecules to their hosts that are sensed by host receptors and that alter basic host physiology, including signal transduction, the cytoskeleton, programmed cell death, and endocytic trafficking. Any one of these steps could be the target of a small molecule that would tip the balance against an infectious microbe.

The Problem

Can genomics help reveal the molecules involved in these steps of pathogenesis? Bacterial sensors can involve histidine kinases, cyclic AMP or cyclic di-GMP, proteins or toxins with enzymatic activity, and glycolipids and lipoproteins that trigger innate immune receptors. Comparative genomics of isolates with differing courses, of related pathogens with differing host range, or of commensals could help define the critical sensors and other genes critical for pathogenesis. Expression analysis of pathogens upon exposure to hosts could help define the critical modifications of the pathogen as it enters the host environment. Protein arrays, comparative proteomics using mass spectrometry, or two-hybrid systems might define ligands and protein interactions.

Differences in host response to pathogens due to host genetic variation might also reveal critical processes in pathogenesis. Resistance might come at any of many levels. Are there populations with increased risk or resistance for classes of pathogens? These human variants could provide valuable insights into the infectious process. Analysis of critical host targets of microbial molecules in human or animal populations may provide evidence for natural selection of these targets and indicate alleles, which promote variability in the response to infectious diseases.

The group will consider these and other more novel approaches where genomics and host resistance may intersect across the full spectrum of microbial pathogenesis.

Initial References

Fauci, A. S., N. A. Touchette, and G. K. Folkers. 2005. Emerging infectious diseases: a 10-year perspective from the National Institute of Allergy and Infectious Diseases. Emerging Infectious Diseases 11(4):519-525.

Rappuoli, R. 2004. From Pasteur to genomics: progress and challenges in infectious diseases. Nature Medicine 10(11):1177-1185.

Relman, D., and S. Falkow. 2001. The meaning and impact of the human genome sequence for microbiology. Trends in Microbiology 9(5):206-208.

Scarselli, M., M. M. Giuliani, J. Adu-Bobie, M. Pizza, and R. Rappuoli. 2005. The impact of genomics on vaccine design. Trends in Biotechnology 23(2):84-91.

Segal, S., and A. V. Hill. 2003. Genetic susceptibility to infectious disease. Trends in Microbiology 11(9):445-448.

WORKING GROUP SUMMARY

Summary written by:

Allison Loudermilk, Graduate Student, Grady College of Journalism and Mass Communication, University of Georgia

Working group members:

- Lawrence Brody, Senior Investigator, Genome Technology Branch, National Human Genome Research Institute

- Brad Cookson, Associate Professor, Laboratory Medicine and Microbiology, University of Washington
- Daniel A. Fletcher, Assistant Professor, Bioengineering, University of California, Berkeley
- Sonja Gerrard, Assistant Professor, Epidemiology, University of Michigan
- Michael Lorenz, Assistant Professor, Microbiology and Molecular Genetics, The University of Texas Health Science Center
- Allison Loudermilk, Graduate Student, Grady College of Journalism and Mass Communication, University of Georgia
- Susan Okie, Contributing Editor, New England Journal of Medicine
- Marc Orbach, Associate Professor, Division of Plant Pathology, University of Arizona
- Mona Singh, Assistant Professor, Computer Science and Lewis-Sigler Institute for Integrative Genomics, Princeton University
- Shankar Subramaniam, Professor, Bioengineering, University of California, San Diego
- Timothy Umland, Structural Biology, Hauptman-Woodward Medical Research Institute

SUMMARY

In November 2005 more than 100 leaders from far-flung fields of science traveled to the National Academies Beckman Center in Irvine, California, to discuss genomics and its enormous implications for infectious disease. The interdisciplinary group hoped to determine whether microbial pathogens—a huge category that spans bacteria, viruses, and fungi—attack host organisms similarly. The answer? Yes.

Day 1: Host or pathogen, the genomic chicken or egg problem

Seated around a rectangular table in Southern California, the members of the working group fell into two camps: those most interested in the host and those expert in pathogens. The informal division echoed that of the greater scientific community, members of which have fanned out to research the common virulence factors of pathogens or the nature of the host response to infectious disease. Both parties have benefited tremendously from rapid whole-genome sequencing made possible by automation. Both are equally important to gaining a better understanding of pathogenesis.

This last fact convinced the group to tackle the assignment from the angles of host and pathogen.

Despite the group struggling to find commonality among endlessly diverse pathogens, several concrete ideas began to form. Studying the nutritional needs of infectious agents, particularly their appetite for iron, was the first idea put forward. If most pathogens require this nutrient, would it be feasible to temporarily deprive the host of iron and kill the bugs, or at least "turn the rheostat toward the host?" Could chelation therapy, a controversial treatment that removes metals from the body, aid in this approach? Regardless of the answer, the group had successfully identified a common need among infectious agents.

The discussion soon migrated to viruses. This formidable class of pathogens is often inhibited by interferon, a key family of proteins that the human body uses to fight viruses on the cellular level and that the pharmaceutical industry synthesizes for antiviral therapies. Many viruses have adapted to evade interferon immune signaling or suppress its production. The catch is that different viruses employ different strategies. For example, one strain of the hepatitis C virus disarms interferon by simulating one of interferon's molecular targets; therefore, despite viruses sharing a similar mechanism, scientists couldn't simply target one common step of viral pathogenesis.

Tampering with the gene expression and replication functions of viruses posed another possibility. When viruses invade cells, they start replicating; therefore, halting this process would prevent the pathogen from gathering strength. Biotech companies have enjoyed some success stopping viral replication by developing compounds that block viral RNA.

Turning to the host for inspiration, the group considered pathogens' preferred mode of entry. The human body is home to several major microbial niches: the mouth, the intestinal tract, the skin, and in females, the vagina. These niches serve as common gateways to the host. The group thought skin cells are the most likely point of entry that can be fortified against microbial attack.

A second idea from the host vantage point was to monitor humans for clues in the form of Toll-like receptors and cytokines. Both types of proteins play a central role in the host's immune response. Toll-like receptors serve as the trigger, sensing the presence of microbes and signaling the immune system to take action. Cytokines follow close on their heels. These regulatory molecules are one of the first types of proteins the host releases after the immune system registers a threat.

Stanford scholar and researcher David Relman's conference tutorial on human microbial pathogens and commensals (or organisms that take part in a symbiotic relationship, in this instance, humans and the microbial flora found in their intestines) spurred the group's final idea. The relationship between host microbiota and infectious disease fascinated the team. Are people more susceptible to disease in the absence of natural microbiota? How do different circumstances alter the microbiota concentration in the host system? Could scientists identify the "good microbiota cops" to give to patients? All were good questions with no ready answers or clinical evidence.

Day 2: Science fiction, cytokines, and Saran Wrap

Two more ideas were discussed as the now familiar team reassembled. Taking a cue from science fiction, one group member wondered whether a cell flush with MHC-compatible inhibitors might be possible. MHC, or major histocompatiblity complex, refers to the set of human genes that code for antigens located on the surfaces of cells. His second idea found a more receptive audience. Building on the earlier discussion of cytokines, he proposed measuring these molecules in the blood and feces of the host, a sort of rapid detection system for infection. The trick was that the baseline level of these chemical messengers would differ from host to host, so healthcare providers would need this patient data on medical records before prescribing care. The other problem is that cytokines are ambiguous by nature; they indicate infection without identifying disease.

The group backed the cytokine idea with a small change. Instead of sampling blood and feces, why not take biopsies from infected patients? If biopsies were not an option, maybe good sources of secondary data existed, such as the National Institute of General Medical Sciences grant awarded to a group of scientists to study how the body reacts to injury on a molecular level.

Setting aside cytokines, the group members rehashed the concept of blocking potential points of entry. What they needed was the medical equivalent of Saran Wrap to preempt invasion. Even a semipermeable, short-term, synthetic barrier akin to Gore-Tex in theory would work. The only snag is that doctors might have a tough time persuading patients to wrap themselves in Gore-Tex or Saran Wrap.

With the clock ticking, the working group switched strategies. Rather than concentrate on the point of entry, the group decided to focus on the class of organisms it wanted to screen against. The team drafted the following outline of its kingdom approach for pathogens:

1. Conduct whole-genome-based comparisons by means of computer analysis for a class of organisms, such as fungi.
2. Identify those genes common to a class of organisms. For example, about 230 genes are present in fungi but not in humans or mice. In addition, many of these genes code for metabolic enzymes and exist as homologues, or similar genes in bacteria.
3. Filter out human genes to identify genes unique to organisms (in this example, fungi).
4. Analyze gene sets to determine whether any have been linked to virulence.
5. Knock out the selected gene in the pathogen and evaluate growth of mutant in vitro (culture) and in vivo (small animal).
6. Develop an in vitro assay for function of gene product.
7. Carry out high-throughput screening, utilizing the assay for gene product function.
8. Develop individual and cocktail drugs.

The next steps would be to modify this eight-step approach to include viruses and develop a viable plan from the host perspective.

Day 3: Finishing the task of innovative science

The final working group session opened with exciting news. Armed with a laptop and the extraordinarily broad, publicly available databases of genome sequencing projects, a group member uncovered some excellent protein targets for drug therapy the previous evening. These targets included HisG in bacteria, His1 in fungi, and others. Even better, the group thought that if it could find nonhuman proteins it could inhibit in fungi, it could apply that same technique to other classes of pathogens, including viruses.

To complete the assigned task, the group still had to return to the other party in host-pathogen interactions, namely, the host. This part was slightly hazier. Some group members did not see the utility of genomics to tailor therapy to individuals. They were more intrigued by the thought of treating patients with nonspecific immunotherapies, such as interferon, that prime the immune system in a general manner, or conversely, use pharmacological agents to interrupt these pathways (this approach would rely on genomics-type information to make diagnoses and institute appropriate treatment). The problem would be knowing when to dispense these therapies, or "when to apply the gas and when to apply the brake."

The practical implication of this proposal was that healthcare providers would need a small collection of patient readouts that measured items indicative of infection at elevated levels, such as cytokines and signaling events from Toll-like receptor stimulation. If the patients' readouts mirrored profile 1, for example, they would receive treatment X. If the readouts resembled profile 2 (or however many profiles were deemed necessary), they would receive treatment Y. The idea was that the universe of pathogens might be infinite, but the number of host responses and treatment options might not be. Richard Jenner and Richard Young's 2005 *Nature Reviews* article, which catalogued 77 different host-pathogen interactions and identified a common host-transcriptional response, also provided some support for the group's view that hosts react similarly to a wide array of pathogens.

Discovering individual thresholds for and responses to infection constituted the most difficult task of the group's host strategy and would entail the collection of massive datasets involving many different populations. One group member suggested that they could collect these data by obtaining informed consent from emergency room patients and defining where these patients were in the course of a disease (retrospectively) and cross-correlate the genomics data with clinical care. Large datasets could be gathered from individual patients, provided that genomic-level observations could be made on single cells. To this end, the National Human Genome Research Institute awarded a grant to develop these technologies. Eventually these datasets could benefit doctors prescribing treatment in individual or pandemic settings.

Just as with pathogen-oriented strategies, the virtue of this approach is that by gathering genomic-level information, the group anticipated the identification of molecular pathways associated with diverse host cell functions, some of which could be modulated to favor quality outcomes for the infected host. Previously known, immunological pathways are one example of potential therapeutic targets. However, the group hypothesized that genomic-level information would provide additional therapeutic targets not necessarily related to immune responses, such as nutritional pathways or cytoskeletal elements. These would be defined as relevant through validation studies, and by definition, the relevance of these pathways to infectious diseases may have been unknown prior to gathering these kinds of datasets.

The other related but less developed host-centric idea was to prevent pathogen invasion by strengthening the host's natural mechanisms for preventing infection—barriers and clearance. Speeding up clearance of waste products and other materials through the gastrointestinal track would mean that pathogens would have less time to establish themselves. Only half-serious, the team thought that the clearance strategy might require more restrooms.

In their short time together, the members of the working group developed testable and workable strategies for blocking the pathways of attack shared by different pathogens. Although the kingdom approach for pathogens may prove more fruitful initially, optimizing the host response to infectious microbes through better therapies and other intervention could change patient care. The group acknowledged that its models would not address all diseases, patients, or pathogens, but they were a start. Now it was time for the greater scientific community to pick up the ideas developed by this interdisciplinary team and build on them.

Explore the Emerging Role of Public Health in Integrating Genomics in Surveillance, Outbreak Investigations, and Control and Prevention of Infectious Diseases

WORKING GROUP DESCRIPTION

Background

According to the 2005 Institute of Medicine workshop report on the implications of genomics for public health, "public health genomics" can be defined as "an emerging field that assesses the impact of genes and their interaction with behavior, diet and the environment on the population's health." The priorities for this field are to:

1. Accumulate data on the relationships between genetic traits and diseases across populations
2. Use this information to develop strategies to promote health and prevent disease in populations
3. Target and evaluate population-based interventions

The "public health system," which includes federal agencies such as the Centers for Disease Control and Prevention, state health departments, and academic public health institutions, is beginning to work closely with basic scientists, professional organizations, consumer groups, and the private sector to "translate" advances in genomics into actions to prevent and control infectious diseases at the population level (Centers for Disease Control and Prevention, n.d.). Increasingly, genetic information from pathogens, the human hosts, as well as vectors will be used to understand the pathogenecity, natural history, and genetic susceptibility to infectious agents. These new types of data could have profound influences on how the public health system conducts its surveillance functions, acute outbreak investigations (Lingappa and Lindegren, 2003), and community-level programs for targeting interventions, such as vaccines and medications. A major challenge is how to apply this information on the population level to affect reduction of the burden of infectious diseases in communities. Current public health education has not fully integrated genomics into its basic competences and core curricula. The practicing public health workforce is not adequately prepared to meet the genomics challenge (Institute of Medicine, 2003; Shortell et al., 2004).

The Problem

What should public health systems do to prepare and respond to the emergence of genomic tools in infectious diseases, in terms of surveillance, outbreak investigations, developing and deploying new interventions (e.g., vaccines), and in its efforts to control of infectious diseases, including bioterrorism events, at the population level?

1. Consider how public health systems should incorporate genomics into acute public health investigations such as outbreak response. What should be the current priorities? Because host genomic factors are involved in determining who will be sick from infectious agents, should public health systems routinely collect such information in their investigations? Should they develop biologic specimen repositories involving pathology tissues, human DNA, etc., to explore the host response to infectious agents and gene and protein expression profiles?

2. Consider how public health systems should integrate genomics into surveillance functions for infectious disease occurrence and tracking in the population. While it may be easier to consider pathogen genomics in surveillance, is there a role for routine collection of human genetic information in such data collection? What tools are needed to make epidemiologic surveillance efforts more in real time?

3. Consider the ethical, legal, and social implications of integrating genomics into public health surveillance and response (e.g., privacy and confidentiality, informed consent) and provide recommendations for action and policy change.

4. Consider the role of genomics in developing and evaluating community interventions for the control of infectious diseases. Such programs include administration of vaccines and working with communities and providers to implement control and prevention measures. Consider the social factors that also play a role in who gets sick from infectious agents (e.g., poor nutrition because of low socioeconomic status could influence one's immunity). What should public health systems do to interact with and educate the public and the provider communities in genomics?

5. Consider the traditional public health data collection categories (e.g., race and ethnicity). Recent articles have underscored one of the consequences of the mapping of the human genome by calling into question our traditional notions of race. Consider how we should categorize individuals given that traditional notions of race and ethnicity are being challenged.

Initial References

Bamshad, M. 2005. Genetic influences on health: does race matter. Journal of the American Medical Association 294:937-946.

Burris, S., L. O. Gostin, and D. Tress. Public health surveillance of genetic information: ethical and legal responses to social risk (pp. 527-548). Online at www.cdc.gov/genomics/info/books/21stcent5.htm#Chapter27, accessed 2/2/2006.

Centers for Disease Control and Prevention. n.d. Genomics and Disease Prevention website at www.cdc.gov/genomics, accessed 2/2/2006.

Institute of Medicine. 2002. *The Future of the Public's Health in the 21st Century.* Washington, D.C.: The National Academies Press.

Institute of Medicine. 2003. *Who Will Keep the Public Healthy: Educating Public Health Professionals in the 21st Century.* Washington, D.C.: The National Academies Press.

Institute of Medicine. 2005. *Implications of Genomics for Public Health: Workshop Summary.* Washington, D.C.: The National Academies Press.

Khoury Muin, J., W. Burke, and E. J. Thomson, eds. 2000. *Genetics and Public Health in the 21st Century.* New York: Oxford University Press.

Khoury Muin, J., W. Burke, and E. J. Thomson. 2000. Genetics and public health: a framework for the integration of human genetics into public health practice (pp. 3-24). Online at www.cdc.gov/genomics/info/books/21stcent1.htm#Chapter1, accessed 2/2/2006.

Lingappa, J., and M. L. Lindegren. 2003. Genomics and acute public health investigations. Genomics and Population Health, United States. Online at www.cdc.gov/genomics/activities/ogdp/2003/chap02.htm, accessed 2/2/2006.

Shortell, S. M., E. M. Weist, M. S. Sow, A. Foster and R. Tahir. 2004. Implementing the Institute of Medicine's recommended curriculum content in schools of public health: a baseline assessment. American Journal of Public Health 94:1671-1674.

WORKING GROUP SUMMARY – GROUP 1

Summary written by:

Lenette Golding, Graduate Student, Grady College of Journalism and Mass Communication, University of Georgia

Focus group members:

- Benjamin Bates, Assistant Professor, Communication Studies, Ohio University
- Sally Blower, Professor, Biomathematics, University of California, Los Angeles
- Karen Burg, Hunter Chair and Professor, Bioengineering, Clemson University
- Robert Cook-Deegan, Director, Center for Genome Ethics, Law and Policy, Institute for Genome Sciences and Policy, Duke University
- Lenette Golding, Graduate Student, Grady College of Journalism and Mass Communication, University of Georgia
- Isaac Mwase, Associate Professor of Philosophy and Bioethics, National Center for Bioethics in Research and Healthcare, Tuskegee University
- Claire Panosian, Professor of Medicine, Division of Infectious Diseases, David Geffen School of Medicine at the University of California, Los Angeles
- Mary Reichler, Centers for Disease Control and Prevention
- Charles Rotimi, Professor and Director, National Human Genome Center, Howard University
- Daniel Salsbury, Managing Editor, Proceedings of the National Academy of Sciences
- Todd Thorsen, Assistant Professor, Mechanical Engineering , Massachusetts Institute of Technology

Summary

This multidisciplinary group was given the task of exploring the interplay between public health and genomics. Recognizing the wealth of information and opportunities that genomics offers public health, the group discussed ways of bridging the gap between gene discovery and the application of genomic tools to surveillance, outbreak investigation, and prevention and control of infectious diseases. These issues were considered broadly, in the context of politics, economics, ethics, race, culture, religion, and the practical limitations of technology and cost.

The first of four working group sessions opened with a discussion about whether recommendations should emphasize tools to detect pathogens or technologies for studying the response of hosts, namely, human beings at risk of catching a disease. Pathogen-centered genomic tools could aid development of rapid diagnostic tests to identify where a pathogen or outbreak originated and whether organisms are drug resistant. Host-oriented genomic tools could be used to assess individual susceptibility to infection and predict response to immunization or specific treatment regimes. Members concurred that genomic tools could broaden scientific understanding of both elements in the host-pathogen equation; they also agreed that neither approach is free of complications.

Lessons from the SARS epidemic

In 2002 an outbreak of a new disease, now called SARS (severe acute respiratory syndrome), began in Hong Kong and was quickly and inadvertently spread by infected people traveling by air. SARS had killed more than 700 people by July 2003. The World Health Organization (WHO) swiftly launched a counterattack against this novel, highly lethal disease. In just seven weeks researchers identified the pathogen and traced it back to its source. In an article that appeared in *Reason*, a nonpartisan political magazine, in April 2003, *Reason's* science correspondent Ronald Bailey wrote, "Thank goodness that SARS broke out in the Genomics Age." Perhaps if the SARS epidemic had occurred in 1992 instead of 2002, many more would have perished.

The group discussed the SARS epidemic in detail, because it illustrates classic public health issues, such as the fact that prompt identification and containment of any pathogen requires collaboration among government agencies and institutions and cooperation from the general public. In addition, the SARS experience reveals the value of such analytical tools as microarrays and computational genetics. The first step in battling any infectious disease outbreak is identifying the pathogen and its mode of transmission. In the case of SARS, new genomic knowledge enabled scientists to identify the viral culprit with stunning speed. This would not have occurred without WHO's excellent coordination efforts and collaboration among the key players in many countries. Driving this effort was public perception—shaped by massive, worldwide media coverage—that SARS was a terrible, lethal, and highly contagious airborne disease that needed to be stopped quickly.

This does not mean that the global public health response to SARS was perfect. A regrettable oversight, according to members of the working group, was the failure to gather information about host genomic factors and susceptibility to SARS from the start of the epidemic. Nonetheless, the group felt that the SARS epidemic of 2002-2003 increased awareness about the importance of good communication, coordination, infrastructure, and surveillance in running effective public health interventions.

Universal issues in the use of genomic technologies: Strategies and solutions

Whether the issue is an infectious disease outbreak or predictive testing for a heritable disease, the group identified five issues that genomic researchers and the public health community must face if genomic tools are to be used effectively. These are:

1. Educating the public
2. Identifying racial and ethnic issues in research
3. Building research capacity
4. Resolving legal and ethical issues around research
5. Targeting research to resource-poor versus resource-rich areas

After much discussion, the group decided to concentrate on two of these: public education and coming to terms with racial and ethnic issues in research.

Educating the public: What is the best way to explain the benefits of genomic research to society?

Scientific advances in the laboratory mean little if powerful officials, or a significant percentage of the public, oppose their use; consider how stem cell research has been slowed by public debate and opposition. Although development and use of genomic tools has not dominated the front pages like stem cell research, there is no doubt that some people mistrust biomedical research in general and are especially nervous that information about their own genetic makeup might be misused in various ways, including denial of health insurance or employment if it were found out that they had "bad" genes. These attitudes and concerns must be addressed, group members said, for genomics to be used effectively against global infectious disease. Moreover, the group felt that the public needs to be educated as to what a genome actually is.

Public campaigns aimed at building acceptance for genomic analysis of individual DNA samples should promote the social benefits of such research. Most people understand that donating blood in times of crisis is an altruistic gesture of real value to others. The group proposed that genomic research programs—where the results could speed drug or vaccine development, for example—could be marketed as an individual's opportunity to do something good for society. It is just as important to communicate the social context of such efforts as it is to explain the facts about what genomic tests do and how new genomic tools work.

The media and entertainment industry could play a huge role in supporting, or hindering, genomics. Entertainment education is a strategy that public health systems have used for years. It consists of intentionally placing educational content in entertainment media. Often the media and entertainment industry can be persuasive in terms of behavior in ways that other advocacy methods cannot. The entertainment education strategy has been utilized successfully for numerous public health campaigns, such as HIV/AIDS prevention and for the promotion of national immunization days.

Incorporation of genomics modules into K-12 education, according to state and national science standards, will also be key to instilling long-term awareness and understanding of the relevant issues and developing a knowledgeable public. A campaign to promote acceptance of genomic testing will need to be ongoing as the field grows and advances are made. To be the most effective, genomic campaigns will need to address regional concerns, local health issues, and language differences; in fact, the individuals directing the campaign will first need to be educated about the campaign target regions in order to best serve the local public. "Buy in" from community leaders will be essential.

Racial and ethnic issues in genomic research

Every individual human genome is a history book. If each were read from cover to cover we would discover we are all from the same place; we would discover that we are all Africans beneath our skin. This is true whether our ancestors later went to Northern Europe and where genes for digesting lactose were selected for, or remained in Africa and developed resistance to malaria parasites. One group member pointed out that group identity is often confused with group ancestry. Instead of being obsessed with relatively minor differences that set apart ethnic groups, the group felt that spreading the message of common origin would help the public understand the importance of global solidarity in the war against disease. But this is a tricky message to convey, because social definitions of race and ethnicity are often confused with genetic origin.

In reality, human genetic makeup is a mosaic, and no gene is unique to one population. Most genetic variation is found in all populations, and occurs among individuals with only a small percentage of variation occurring at the population level. Social and ethnic conceptions of "race" are commonly understood only at the population level. This is why the notion of race is often not necessarily helpful in understanding disease. Some genes that have health effects do, however, occur more commonly in some populations, and correlate roughly with ancestry where such ancestry maps to genetically distinguishable populations. Some health-related factors may, therefore, correlate with "race" or "ethnicity," although we should understand that the categories are only rough proxies for underlying biological differences. Most "racial" differences are likely to be affected by social, economic, or environmental factors at least as much as genetic differences, but some differences will correlate with genetic differences, and the tools for finding those genetic differences have advanced considerably in the past decade. The important issue here is not race itself but the genes that predispose a person to disease.

The group reflected on BiDil, an antihypertensive drug to treat congestive heart failure that the Food and Drug Administration (FDA) recently approved specifically for use in self-identified African-American patients. This is the first drug approved for use on a racial basis, and the FDA took this step after this medication failed to show efficacy in a large, mixed-race sample but performed significantly better in a smaller follow-up trial restricted to a self-identified group of African Americans. No genomic data were collected in the trials; instead, race—a much cruder and self-selected "identifier"—was used as the inclusion criterion for the follow-up trials.

As a consequence, the group agreed that it is impossible to determine whether race, ethnicity, or environmental factors explain why BiDil appears to work better in some people than others. More to the point, if the underlying differences are biological, attributable to gene frequency differences, then only some African populations (and hence African Americans) would have the high-frequency alleles and others would not. It is extremely unlikely, for example, that those from sub-Saharan Africa and those from northwestern Africa and those from northeastern Africa would all have the same allele frequencies, given the high level of diversity within Africa. It is also highly likely that some non-African populations also have varying allele frequencies, and so some individuals within those populations could benefit. By using a social measure, such as self-identified race, without the underlying data about genetic variations, the story cannot be understood, and clinical decisions are based on a rough heuristic. This is unfortunate when we have in hand the technologies to do the genomic analysis that could sort out the causal pathway. Overall, the group felt the BiDil study represented a missed opportunity to correlate genomic information with drug efficacy.

The story of race and BiDil is a cautionary tale for host differences in response to pathogens. There are apt to be population differences, particularly for pathogens that have co-adapted in particular regions over long periods of time, so there has been selection pressure on both pathogens and human hosts that may well map to geographic areas and environmental factors that influence the prevalence of specific infectious diseases. It would be sad, indeed, if the story stopped at "race" when in fact, the underlying story is specific environmental factors, host factors, or pathogen differences that could be illuminated by genomic tools.

Benefits to society

The group felt strongly that if the wealth of information stored in human genomes is going to be harvested for the good of humankind, then social, political, and economic barriers and global collaborations must be addressed. The intersection of patients, government, politics, and genomics is often a cacophonous and confusing place. The full potential of the genomic revolution will not be easily realized. The exchange of genomic information across international borders involves delicate political negotiations as well as vast expenditures of capital, while guaranteeing individual rights and ensuring that data will be used for good, not evil, purposes. The exchange of genomic information will require leadership and will mandate education of the leadership such that local traditions and customs are respected and local "buy in" is ensured. These challenges are well worth tackling, however. Stepping back to look at the big picture, the group emphasized that the true promise of genomic technology is saving lives and preventing human misery in all the nations of the world.

WORKING GROUP SUMMARY – GROUP 2

Summary written by:

Corey Binns, Graduate Student, Science Journalism, New York University

Focus group members:

- Corey Binns, Graduate Student, Science Journalism, New York University
- Ronald W. Davis, Professor of Biochemistry and Genetics and Director, Stanford Genome Technology Center, Stanford University School of Medicine
- Georgia M. Dunston, Professor, Microbiology, National Human Genome Center, Howard University
- Stephanie Malia Fullerton, Assistant Professor of Medical History and Ethics, University of Washington School of Medicine
- Lyla M. Hernandez, Senior Program Officer, Board on Health Sciences Policy, Institute of Medicine
- Ezra C. Holston, Assistant Professor, School of Nursing, University of California, Los Angeles
- Barbara R. Jasny, Supervisory Senior Editor, Science
- Rima F. Khabbaz, Acting Deputy Director, National Center for Infectious Diseases
- Hod Lipson, Assistant Professor, Mechanical and Aerospace Engineering, Cornell University
- Daniel Oerther, Associate Professor, Civil and Environmental Engineering, University of Cincinnati
- Anne W. Rimoin, Assistant Professor of Epidemiology, University of California, Los Angeles, School of Public Health
- Marc S. Williams, Director, Clinical Genetics Institute, Intermountain Healthcare

Summary

Back in 1918 it could take weeks to travel from one country to another, and yet the Spanish flu pandemic claimed more than 20 million lives as it spread around the globe. Today one can travel halfway around the world in a matter of hours and disease can spread farther and faster than ever before. Fast-paced international travel and trade have brought the world closer together. This new intimacy has brought infectious disease to the forefront of global public health issues. Thus, infectious disease control is now a shared burden—as a growing threat in developed nations, and a clear and present danger in developing nations—and it is therefore a perfect place for a new science like genomics to play an important role. The need to develop new approaches to fight infectious disease was evident in the sentiments of scientists, engineers, and medical researchers gathered in Irvine, California, at the Third Annual National Academies Keck *Futures Initiative* Conference.

Although the developed world focuses much of its medical research on the diagnosis and treatment of chronic diseases, the developing world is constantly compromised by infectious disease. In these countries, people still suffer from a lack of basic care, such as unclean drinking water, malnutrition, and poor sanitation.

Genomics is a discovery science whose specific applications have yet to be completely realized. While it is difficult to predict exactly how much benefit will result from investing in this kind of research, the time is ripe for public health to begin to explore its potential contribution to mitigating illness and death from infectious diseases. Thanks to the Human Genome Project and enthusiastic news coverage, genomics has received a lot of public attention. But celebrity can result in mistaking genomics for the answer to just about everything in health care, and proponents risk overselling its potential. Aware of these dangers, the 12 members of the group discussed the future of genomics in the battle against infectious disease.

The group agreed that it is essential to set priorities with respect to how best to proceed in applying innovative science in the fight against infectious disease. While innovative science is essential to advance the knowledge base for public health, resource limitation is an omnipresent barrier to improving public health. The investment in genomic science has to be balanced against opportunities for implementation of programs that use inexpensive intervention. For example, a simple bed net to protect against malaria-infected mosquitoes may be just as effective, and far cheaper, than high-tech treatments based in genomic science. Investment in genomic solutions must therefore be viewed through the long-term lens of the opportunity cost—that is, what will we not be able to afford to do if we spend resources on genomics, and what benefits are likely to accrue in the future.

To help understand the etiology and pathogenesis of a disease, determining the genetic sequence of a pathogen might be a higher priority than sequencing the human hosts, because sequencing the human genome costs are enormous and may yield less information about the infectious disease. While resistance and susceptibility factors in the human genome will no doubt prove important in the future, the price of such discovery has to be balanced against more pressing public health needs. In addition, diseases carrying the highest burdens should be given the highest priority for such research, so that the public benefit to scientific discoveries is possible. Time and resources should be allotted appropriately so that these diseases, namely AIDS, malaria, and tuberculosis, are studied first. Resources should also be given to apply genomics to further research and development of new vaccines and medications because vaccines have historically topped the list of groundbreaking improvements made to public health. Finally, a focus should also be placed on identifying and characterizing specific disease-causing organisms in order to break the chain of disease transmission. If genomics can facilitate this activity, then the expenditure of resources will be justified.

It is clear that genomics fits into the circle of public health. It is important, and not too early, to invest in genomics research—a balanced public health portfolio should include allocating and utilizing resources in discovery science. Analysis of large and diverse kinds of data will be crucial. For instance, host genomic research may lead to a better understanding of susceptibility to infections and risk for disease, as well as the beneficial and adverse effects of drugs and vaccines. The group supported the idea of a bioresource—a place where quality-controlled information could be collected, analyzed, and applied to diagnose, treat, and prevent disease.

Determining exactly what goes into the bioresource was one of the first matters to tackle. In addition to information that is traditionally collected (such as epidemiologic and demographic data), biosamples, ecological information, and genotypic information on pathogens and hosts should also be compiled. The locations of collection and storage sites must be geographically distributed to ensure that participating countries and organizations are engaged and share ownership and responsibility in maintaining the bioresource—a model of data and information gathering that contrasts with the more "colonial" model of data reposing only in the first world. Designers of the bioresource must place a strong emphasis on standardizing the information collected and enforcing quality control. This will ensure that researchers and healthcare professionals who use the bioresource for diagnoses are working with accurate, standardized datasets.

The group also agreed that the data that come out of the bioresource are as important as the resources devoted to it. Therefore, the resource must be managed as an intelligent system that learns and adapts, continually improving its ability to predict and diagnose health problems as more information is collected in the system. As the study of genomics matures and more discoveries are made, the bioresource would be built to continually adjust to these new insights with new applications. Finally, and key to its success, the bioresource's output of information has to be almost universally accessible: easy to use by anyone from anywhere.

The benefits of such a bioresource could be far-reaching, not only as a research tool but also as a device to improve public health. Programs for collecting samples and information for single diseases exist: preexisting HIV and malaria projects act as good starting points from which to develop a larger resource that covers multiple diseases. An excellent model is the Global AIDS Program, which partners with communities, scientists, and public officials to prioritize health concerns and direct prevention programs appropriately. However, the Global AIDS Program funding is specifically allocated for prevention programs, and not for research, which would be a crucial additional benefit of the proposed bioresource.

At the same time, it is important to remember that genomics can be successful only if the scientific community takes the time and energy to teach the public how it will improve their health, because of the inherent need for collaboration between the two groups. The group unanimously agreed that education, in the form of a dialog among all parties, is essential to promoting the use of genomics as a public health tool, especially when scientists need the cooperation and understanding of the public to collect genetic information in the form of blood or saliva samples. The public can in turn teach the scientific community what health issues are of greatest concern to them and offer insight into environmental and social aspects of pathogenicity. Both sides have something to learn from each other, and the process will help everyone build trust in the science and one another.

The public's fears that new technologies may be misused or create inequalities can be put at ease only by taking seriously, and then frankly addressing, their concerns. The public may be opposed to a bioresource because of its potential to be used inappropriately and possible negative implications from collecting genetic information from infected individuals. Discrimination has historically been a complication of communicable disease outbreaks. Characterization of lepers as unclean is but one example of the stigmatization resulting from infectious disease. Genetic identification gives the public cause to worry about another form of discrimination. We hope that by educating both healthcare workers and the public, we can prevent these problems.

The task of winning over the global public is never easy. With the social and economic disparities that already exist between the developing and the developed nations, it is difficult to make public health programs attractive to those who are uncertain of their place in the definition of "public" health. The bioresource could potentially serve as a tool to help engage all participating and interested parties in the collection, storage, analysis, and dissemination of the information. However, it will likely be a challenge to articulate the message so as to compel a strong commitment to investing in this as a shared public health global resource.

The International Human Genome and HapMap project exemplify the inherent challenges of equality in such ambitious global health initiatives. The principle behind the project is to make all information freely available to any scientist in the world. While this open access model appears wholly equitable, many scientists and the public can't use the open access model for human genome sequence and HapMap data because they have neither the training, equipment, nor the funding to make use of the available information. Without a commitment to provide public education and open access to equipment, resources such as the human genome sequence and HapMap (or the proposed bioresource) will have only limited impact, particularly for disenfranchised groups and those in traditionally underserved communities.

In addition to solving the problems of inequality and global accessibility of the resource's assets, there are many ethical concerns that come with a system as encompassing as the group's proposed bioresource. When donating genomic material to a biobank like the bioresource, participants are typically expected to provide consent for the use of such materials in subsequent research. In exchange, investigators pledge to protect confidentiality and assure contributors that their genetic information will remain secure. Safety is of utmost importance with any repository of infectious materials, and steps must be taken to protect the resource from any harmful misuse of its samples. However, individuals will also have compelling interests in knowing the health status of themselves and their families, especially in the cases of treatable diseases.

With genomics still in its infancy, now is the perfect time to concentrate on the challenges and promises of this new science in winning the battle against common infectious diseases in their threat to human life and the public health.

What Will It Take to Sequence an Individual's Genome for Under $1,000 in Less Than 10 Years?

WORKING GROUP DESCRIPTION

Background

Following the sequencing of the human genome, the next challenge is to ascertain how variations among genomes of individual persons play significant roles in their variations in disease susceptibility and severity and the efficacy of medical therapies. An individualized approach to medicine requires the ready acquisition of information about the sequence of individual persons' genomes with reliable technology at acceptable costs. Thus, the ability to sequence an individual person's genome for under $1,000 within the next 10 years has been set as a goal.

The Problem

- What are the critical technological roadblocks and fundamental biological showstoppers? The traditional method of sequencing involves gel-based technology with optical detection. The estimated cost of using such a technology for each entire genome would be approximately $20 million. However, a variety of new sequencing methods is under development. For example, Shendure and colleagues (2005) have developed a new technology that uses color-coded beads (~1 μm in diameter), a microscope, and a camera to replicate thousands of oligo strands of DNA with each strand on its own tiny bead. Fourteen million such beads can be packed in an area of the size of a dime. Each camera frame can analyze beads, each of which has one of four dye colors. The flow of the beads is computer controlled and the camera records the dye color and hence the sequence. It is estimated that the cost per base with this new technology is ~1/9 of conventional sequencing. With continued development of novel technologies (Shendure et al., 2005; Margulies et al., 2005) that can reduce the cost of sequencing by factors of 10, the goal of $1,000 should be achieved. With a reference sequence to guide analysis of individual genomes, assembly is straightforward.

- Is it indeed so far fetched or can it be done for each individual using analysis of the cluster (macro-level) variations from the baseline of a generic human genome? It is expected that the average variation across humans in the genome sequence is 0.1 percent or less, and the changes in the coding regions are much smaller. The single nucleotide polymorphism and haplotype mapping projects are beginning to provide the variations across humans that might give rise to pathology. Initially, efforts to define disease-causing variants will use SNP (single-nucleotide polymorphism) markers, enhanced in power by the HapMap, but this will switch to whole genome analysis as sequencing costs drop. The challenge then will be to identify the causative changes among the many revealed differences. Comparative sequence analysis is rapidly identifying the estimated 5 percent of the genome that is well conserved and appears to be functional, thus substantially narrowing the search.
- What will we do with the information for diseases?

— The awareness of the potential for specified diseases would provide an opportunity for changing lifestyles that would be more conducive to healthy living.

— The awareness would also enable medical monitoring on a periodic basis along with risk assessments (genomic triage).

— In a decade it is likely that gene therapy and other therapeutic methods will begin to take shape for targeted therapeutics. Identification of specific gene defects through selective sequencing can aid in focused efforts for gene therapy.

Initial References

Margulies, M., M. Egholm, W. E. Altman, S. Attiya, J. S. Bader, L. A. Bemben, J. Berka, M. S. Braverman, Y. J. Chen, Z. Chen, S. B. Dewell, L. Du, J. M. Fierro, X. V. Gomes, B. C. Godwin, W. He, S. Helgesen, C. H. Ho, G. P. Irzyk, S. C. Jando, M. L. Alenquer, T. P. Jarvie, K. B. Jirage, J. B. Kim, J. R. Knight, J. R. Lanza, J. H. Leamon, S. M. Lefkowitz, M. Lei, J. Li, K. L. Lohman, H. Lu, V. B. Makhijani, K. E. McDade, M. P. McKenna, E. W. Myers, E. Nickerson, J. R. Nobile, R. Plant, B. P. Puc, M. T. Ronan, G. T. Roth, G. J. Sarkis, J. F. Simons, J. W. Simpson, M. Srinivasan, K. R. Tartaro, A. Tomasz, K. A. Vogt, G. A. Volkmer, S. H. Wang, Y. Wang, M. P. Weiner, P. Yu, R. F. Begley, and J. M. Rothberg. 2005. Genome sequencing in microfabricated high-density picolitre reactors. Nature. Published online before print, Jul. 31. Online at www.nature.com/nature/journal/vaop/ncurrent/abs/nature03959.html, accessed 2/2/2006.

Pevzner, P. A., H. Tang, and G. Tesler. 2004. De novo repeat classification and fragment assembly. Genome Research 14(9):1786-1796. Erratum in Genome Research 14(12):2510.

Shendure, J., G. J. Porreca, N. B. Reppas, X. Lin, J. P. McCutcheon, A. M. Rosenbaum, M. D. Wang, K. Zhang, R. D. Mitra, and G. M. Church. 2005. Accurate multiplex polony sequencing of an evolved bacterial genome. Online at Sciencexpress www.sciencemag.org/cgi/content/abstract/1117389v1, accessed 2/2/2006.

WORKING GROUP SUMMARY

Summary written by:

Leah Moore Eisenstadt, Graduate Student, Science Journalism, Boston University

Working group members:

- Steven Brenner, Associate Professor, Plant and Microbial Biology, University of California, Berkeley
- Siobhan Dolan, Obstetrics, Gynecology and Women's Health, Albert Einstein College of Medicine and March of Dimes Birth Defects Foundation
- Leah Moore Eisenstadt, Graduate Student, Science Journalism, Boston University
- Mark Guyer, Director, Division of Extramural Research, The National Human Genome Research Institute
- Leonid Kruglyak, Professor of Ecology and Evolutionary Biology, Lewis-Sigler Institute for Integrative Genomics, Princeton University

- Babak Parviz, Assistant Professor, Electrical Engineering, University of Washington
- Holger Schmidt, Assistant Professor, Electrical Engineering, University of California, Santa Cruz
- Eric Topol, Provost and Chief Academic Officer, Cleveland Clinic Foundation
- Victor Ugaz, Assistant Professor, Department of Chemical Engineering, Texas A&M University
- Huntington Willard, Director and Professor, Institute for Genome Sciences and Policy, Duke University
- Shaying Zhao, Assistant Professor, Biochemistry and Molecular Biology, and Institute of Bioinformatics, The University of Georgia

Summary

On April 14, 2003, scientists announced that the Human Genome Project was essentially complete—two years ahead of schedule. After 13 years of work, the three billion base pairs of the human genome were known in great detail and with high accuracy. *The Washington Post* said that the project "revealed in exquisite detail the genetic blueprint underlying all human life." *The Boston Globe* deemed the genome "the last milestone for one of the modern era's grandest scientific endeavors."

The project cost $3 billion to complete, and despite today's technology, which has advanced significantly since the start of the project in 1990, it would cost a whopping $20 million to do it again. Lowering the cost of sequencing would make the genome both a useful research tool and a potential avenue to personalized clinical management and treatment. Working group 12, therefore, was given the task of answering the question: What will it take to sequence an individual's genome for under $1,000 in fewer than 10 years?

As the group first gathered in the room, the team of biologists, engineers, and institute directors threw around the idea of what the group's task actually was. Some members interpreted the question as requiring answers of a technical nature, while others saw it as merely a conversation starter to discuss the future of genomic sequencing technology.

Why do we need the $1,000 genome?

One issue that was quickly brought to the table, and that persisted throughout the sessions, was the necessity of a $1,000 genome. A cheaper genome could one day inform patients of their genetic risk for cancer, or it could allow researchers to identify the next avian flu. Because the group's initial task referred to an individual's genome, the question of clinical outcomes was on everyone's mind. The group, however, saw the research uses of the $1,000 genome as being much more promising much sooner. A cheaper genome could allow scientists to characterize all the differences between a given human genome and a reference genome and help identify possible genetic sources of disease. Being able to quickly and very cheaply classify bacterial or viral pathogens in the field would also be a powerful tool.

The usefulness to research is a "no brainer," said Eric Topol, of the Cleveland Clinic Foundation, but the value for clinical practice was another question entirely. Siobhan Dolan, a physician with Albert Einstein College of Medicine and the March of Dimes, agreed: "What do the genome bases mean for health care?" Dolan was concerned with the clinical uses of widely available, low-cost genomic information, and reflected on her experience with newborn baby testing. "You get a printout, give it to the parents, and say, 'This is what you have, go!'" The group also noted that few diseases are monogenetic in origin, which complicates the genotype-phenotype relationship. For diseases that result from the complex interplay of several genetic mutations and epigenetic factors, the path from identifying the mutations in the lab to a treatment in the clinic may not be smooth. Others acknowledged the cart-before-the-horse result of genetic information that comes without the knowledge of what to do with it, but some members of the group suggested leaving sociology out of the discussion.

Even if the $1,000 genome were available, the group wondered, would the clinical outcomes be worth it? Screening for a few relevant SNPs might give a patient as much relevant, clinical information as the whole genome but with a much lower price tag. As the human genome has only been sequenced recently, however, there are potentially important areas of the genome that wouldn't be included in a screen limited to SNPs. "Who can declare what is junk DNA?" asked Hunt Willard, director of the Institute for Genome Sciences and Policy at Duke University, adding, "I'd be afraid to bet my health on it." After discussing the utility of the complete genome versus an incomplete one, the group decided that the cost-effective sequencing of complete genomes would be vital for research in the upcoming years, potentially yielding information about previously overlooked areas of the genome.

What does the $1,000 genome mean?

One step these experts took was to redefine what version of the genome should be reduced to $1,000. Babak Parviz, assistant professor of electrical engineering at the University of Washington, noted that to improve human health, genomes would be needed for both humans and pathogens, both small and large. Some microbes and parasites have genomes of 3 to 25 megabases, while the human genome is 3 gigabases long. Therefore, flexibility in the technology to sequence genomes would be best. Ideally, the cost of sequencing the genome would be scalable, lowering the cost of yeast or bacterial genomes to $0.30 per megabase.

The working group's task was questioned by Leonid Kruglyak, professor of ecology and evolutionary biology at the Lewis-Sigler Institute for Integrative Genomics at Princeton University. "It's not the main goal to resequence genomes over and over," he said, referring to the narrow focus of the working group's question at hand. "The main goal is to get cheaper sequencing in general." Mark Guyer, director of the Division of Extramural Research at The National Human Genome Research Institute, agreed. "Focusing on the $1,000 genome may be too oversimplifying," he said. "What we're talking about is cheap data generation." The group concurred, and chose to redefine the $1,000 genome as a metaphor for improving technology for sequencing and for capturing genomic variation. The improved technology, in turn, would spur reduction in costs for sequencing other genomes, such as the $1 bacterial genome and the $5 parasite genome.

Another point of discussion was the accuracy and completeness of the $1,000 genome. One extreme is a genome that has over 90 percent accuracy over 80 percent of the genome with tens of thousands of false positives and no information about structural variation. The other extreme is letter-perfect sequence with technology that also captures variations in structure, copy number, translocations, and inversions. Thus, the full clinical utility will depend not merely on the $1,000 needed for a complete genome sequence but also the cost of additional technologies to assess genome and chromosome structure (currently assessed by karyotyping), as well as the copy number of each segment of the genome (currently assessed by methods such as array-based comparative genome hybridization). "I want the latter, not the cheapo version," said Eric Topol. "Maybe structural things and epigenetics are important in health and disease; we don't know how much simple sequence is tied to that."

How do we get to the $1,000 genome?

When the Human Genome Project began in 1990, the cost of sequencing was $10 per base pair. The project used sequencing methods based on those developed by Frederick Sanger, which involves polymerase chain reaction (PCR) amplification of DNA that is then separated by gel electrophoresis. During the genome project, the cost of sequencing dropped 2,000-fold, now approaching $0.10 per base pair. In order to get the total cost down to $1,000, the cost needs to drop another 20,000-fold, which is not impossible when compared to some of the other fantastic advances in technology. "Compared to advances in semiconductors," said Babak Parviz, "this is not science fiction." Electronic chips can have billions of devices and sell for $200, but it took almost four decades to increase from 4,000 transistors per device in the 1960s to billions today.

The group discussed the source of the genome's current sticker shock. Shaying Zhao, a University of Georgia molecular biologist, suggested that much of the cost lies in the fluorescent dye used in conventional sequencing. It has to be a revolutionary and radical change, she said, to bring the cost down so many orders of magnitude. "[But] even if you improve the dye to decrease cost," Zhao added, "a $1,000 genome is still impossible with the current methods." Ron Davis, a geneticist and biochemist from Stanford University, who visited the group during the second two-hour session, said that sequencing costs are split evenly between labor, reagents, and instrumentation.

New and exciting technologies that can revolutionize sequencing methods are in development by private companies and academic researchers. Microelectrophoretic devices use microfabricated wafers containing 384-well capillaries. DNA is injected at the perimeter of the wafer and runs toward the center, where detection of the bases is carried out by confocal fluorescence scanning. Another new technology is nanopore sequencing, which lets single-stranded DNA strands pass single-file through a nanopore in a lipid bilayer. With this technology, ionic current through the open channel drops as various polynucleotides pass through the pore, but the pores must be improved to achieve single-base resolution.

Another new method is sequencing by hybridization, in which the differential hybridization of oligonucleotide probes reveals the sequence. In one type of hybridization sequencing, DNA is immobilized and serial hybridizations are carried out with short probes. The differential binding of specific probes to DNA can be used to identify the sequence. In another type of hybridization sequencing, used by the companies Affymetrix and Perlegen, probes are immobilized to arrays of sample DNA. Each array, or chip, has four features, with the middle of each feature being either an A, C, G, or T base pair. Labeled sample DNA is hybridized to the chip and the sequence is determined by measuring which feature gives the strongest signal. Hybridization technology faces challenges in approaching $1,000 for a whole genome: avoiding cross-hybridization of probes to incorrect targets, and the requirement for sample preparation, such as PCR amplification.

Cyclic-array sequencing methods, such as fluorescent in situ sequencing, pyrosequencing, and single-molecule methods, take advantage of the power of parallel sequencing. These methods use repeated cycles of polymerase extension, or synthesis, with one nucleotide at each step. The magnitude of the signal during each cycle can be used to infer the order of bases in each sequence. All cyclical methods involve amplification steps that are spatially isolated. One company, 454 Corporation, scaled up pyrosequencing by using thousands of parallel picoliter-volume PCR amplifications. One difficulty with cyclic-array methods is consecutive runs of the same base. Relative amounts of signal may be used to reveal the length of those runs, but one solution involves reversible terminators that would enable simultaneous use of all four bases. Another cyclic-array method eliminates the need for amplification by directly sequencing single molecules. Some companies developing this technology include Solexa, Genovoxx, Nanofluidics, and Helicos. But our group recognized that significant public and private funding would need to be in place for these companies to develop new technology, in addition to a biological driver.

Mark Guyer told the group that the National Human Genome Research Institute is spending $25 million per year in grants to (1) reduce the cost of sequencing by two orders of magnitude to $100,000 within 5 years and (2) to reduce that to $1,000 over the next 10 years. Ron Davis said, "I think it's technically feasible in 10 years." Robert Waterston, who worked on the first human genome, told the group, "It's going to be done." He noted that Solexa would release a machine next year that it says will give 10-fold coverage of the genome for $100,000. That price tag will only get smaller with time.

The genome isn't everything

If given the capability of a $1,000 genome tomorrow, Hunt Willard said he would spend a decade doing basic and clinical research, and eventually translate that into clinical care. Everyone agreed, knowing the hurdles of using genomic information in the clinic. The group agreed that a cost-effective human genome would not be a panacea in terms of health care. Eric Topol said, "No matter if you had the whole ball of wax, it's only a small piece of the story, not including the interaction of genes or environment." Hunt Willard responded, "But you could do more research faster with the $1,000 genome."

Even when genomic information can inform medical decisions, physicians may be reluctant to use it. "It's the technology/culture divide," Willard said. "If you delivered this information [to clinicians] tomorrow, people would drop it like it's a hot potato; they'd have no idea what to do with it in the current healthcare climate." It's not the healthcare climate that's to blame. The fact is, science doesn't yet know what all the information means. Genomics is not ready for prime time in routine medical practice. But Topol voiced the core of what motivates and inspires researchers to improve sequencing technology: "If you can cut through that divide, the opportunities for understanding the biological basis of disease is extraordinary."

How Can We Use Natural Variation in Disease Resistance to Understand Host Pathogen Interactions and Devise New Therapies?

WORKING GROUP DESCRIPTION

Background

The genetic variation in the genomes of pathogenic microbes and the organisms they infect provides a DNA sequence record of the evolutionary "arms race" between host and pathogen. Specific pathogens are increasingly recognized as powerful selective forces in the evolution of all organisms and similarly the need for the pathogen to adapt to the host is a major force driving pathogen evolution. Dramatic recent progress in genetics and genomics provides numerous exciting insights into this process, which may identify previously unrecognized host defense pathways, as well as new opportunities for therapeutic intervention.

Polymorphic susceptibility and/or resistance alleles at multiple genetic loci have been identified in human populations, as illustrated by the classic example of the sickle cell hemoglobin mutation, which confers resistance to malaria. The spectacular resources now available with the completion of the genome sequences for numerous mammalian hosts, as well as their specific pathogens, provide unprecedented opportunities to dissect these complex pathways of interaction and to identify new targets for therapeutic intervention.

The Problem

• Consider the numerous examples of genetic variation that contribute to the host response to infectious pathogens in terms of resistance, increased susceptibility, or varying response. Are there any general themes that can be derived from the growing number of examples of such genetic variants and are there specific approaches that can be taken at the genome level to identify large numbers of clinically important variants?

• As such specific resistance and susceptibility alleles are identified, often with widely different prevalences among human populations, what specific social or policy issues does this raise when approaching these populations?

• What approaches should be taken to increase the interaction between infectious disease experts and geneticists in harvesting this enormous dataset? Are new database structures and novel bioinformatic approaches necessary to effectively analyze this sequence variation information, including the interaction between the separate but ultimately related genomes of host and pathogen?

Initial References

Dean, M., M. Carrington, and S. J. O'Brien. 2002. Balanced polymorphism selected by genetic versus infectious human disease. Annual Review of Genomics and Human Genetics 3:263-292.

Fortin, A., M. M. Stevenson, and P. Gros. 2002. Susceptibility to malaria as a complex trait: big pressure from a tiny creature. Human Molecular Genetics 11(20):2469-2478.

O'Brien, S. J., and G. W. Nelson. 2004. Human genes that limit AIDS. Nature Genetics 36(6):565-574.

Segal, S., and A. V. Hill. 2003. Genetic susceptibility to infectious disease. Trends in Microbiology 11(9):445-448.

WORKING GROUP SUMMARY – GROUP 1

Summary written by:

Aria Pearson, Graduate Student, Science Writing, University of California, Santa Cruz

Working group members:

• Philip Awadalla, Assistant Professor, Genetics, North Carolina State University

- James Brody, Assistant Professor, Biomedical Engineering, University of California, Irvine
- Elliott Crouser, Assistant Professor, Division of Pulmonary and Critical Care Medicine, The Ohio State University Medical Center
- Alison Farrell, Senior Editor, Nature Medicine
- Aria Pearson, Graduate Student, Science Writing, University of California, Santa Cruz
- Christopher Plowe, Professor, Medicine, University of Maryland School of Medicine
- Bernhard Rupp, Structural Genomics Group Leader, Biosciences, University of California – Lawrence Livermore National Lab
- Katherine Spindler, Professor, Microbiology and Immunology, University of Michigan Medical School
- Sarah Tishkoff, Associate Professor, Biology, University of Maryland
- Elizabeth Winzeler, Associate Professor, Cell Biology, The Scripps Research Institute
- Hsiang-Yu Yuan, Graduate Student, Institute of Biomedical Sciences, Academia Sinica
- Hongyu Zhao, Ira V. Hiscock Associate Professor, Public Health and Genetics, Yale University
- Michael Zwick, Assistant Professor, Human Genetics, Emory University School Of Medicine

Summary

Two people catch the same common cold. One suffers from a raging sore throat, weeks of sniffling and sneezing, and a nasty cough. The other breezes through it, barely noticing a sniffle or two.

Why? Is sufferer number 2 blessed with genetic differences that confer natural resistance? How can we use this natural variation to illuminate host-pathogen interactions and pave the way for new therapies?

Asking these questions may seem silly when talking about the common cold, but when dealing with diseases such as malaria, TB, or HIV, it could mean life or death.

At the end of four intense discussion sessions spread over three days, the working group—which included some of the leading scientists in genetics, medicine, immunology, cell biology, and engineering—had outlined an ambitious strategy for tackling the problem. But first they had to redefine it.

"We can't leave the pathogen out," one group member said. The group decided their assigned question, "How can we use natural variation in disease resistance to understand host-pathogen interactions and devise new therapies?" implied restricting the discussion to genomic variation in the host. The group felt that in elucidating host-pathogen interactions the pathogen could not be ignored, so the group decided to broaden the topic to include natural variation in the pathogen as well. Then, going one step further, the group decided to throw environmental variation into the mix, because environmental factors are sure to play a significant role in disease resistance.

Examples of natural resistance abound. There's the famous relationship between the allele for sickle cell anemia and malaria resistance. Another example was provided by a group member who described a village in Africa where two tribes live together. Members of one tribe are completely resistant to monkey pox while members of the other tribe die from it. Another group member mentioned hepatitis B. Eight percent of people in China are infected with the virus, but while some die, others are fine. Finding the underlying cause of this variability in disease resistance and using that knowledge to combat disease is becoming increasingly possible, given the advances in genomic science.

The group decided to focus on humans, a marked contrast to the other focus group dealing with this question, which decided to start with mouse models. Every member of the group agreed that sequencing the genomes of all humans would be ideal, however, that being unrealistic, the group thought it sensible to begin with large-scale association studies, most likely conducted in Africa, where the most variation exists and the burden of disease is profound.

The interdisciplinary studies would involve sequencing all host and pathogen genomes to identify the mechanisms of interaction and pinpoint therapeutic or vaccine targets. The ethical and legal issues would be staggering, but the group felt the potential benefits to society of controlling or eradicating deadly diseases would make the endeavor worthwhile.

Of course, the cost of sequencing a human genome (currently around $20 million, though that price tag will soon drop to around $100,000, according to scientists at the conference) will have to come down significantly before such studies are feasible. But technologies are constantly improving and the $1,000 genome is not far off, according to another group at the conference.

Large cohort studies: It's all about the phenotype

One of the key problems that arise when carrying out large-scale association studies is the issue of phenotypic characterization. Diagnosing diseases accurately is difficult in developing countries like Africa, where many people are infected with more than one pathogen, as well as multiple strains of the same pathogen, and environmental conditions are extremely variable.

The best approach is an integrative analysis, in which the host genome and the pathogen genome are evaluated together, in an environmental context, to determine the effects on the phenotype. This would require people who view things from the perspective of the host to work closely with people who analyze the behavior of pathogens, an important development resulting from these types of studies.

"In most medical schools, for example, there's a department of infectious disease and there are people who work in human genetics and they rarely talk to each other," the group spokesperson said at the final presentation. This has slowed progress in genomic research in regard to infectious disease, according to the group. The studies the group proposed would get the two disciplines talking, which is one of the goals of the National Academies Keck *Futures Initiative* Conference.

The studies would also require tools for detecting and analyzing multiple infections, and multiple strains of the same infectious agent—quickly and cheaply. An ideal device would be handheld, would run on batteries, and be able to separate and sequence the host genome and the genomes of all the microbes in a single drop of blood.

Assuming that sequencing becomes affordable, and appropriate technologies become available, another challenge associated with these studies is finding large, diverse study populations. "To really pull this thing off you may need study samples on the size of 10,000 humans," the spokesperson said.

With projects of this scale, the ELSI—or ethical, legal, and social issues—would require serious time and attention. For instance, there would be an obligation to follow up with medical treatment for all the participants. "If you're going to screen for something, you're obligated to tell them about it and treat them," a group member pointed out. Cross-cultural communication issues would also arise. The group felt that to alleviate some of these problems, partnerships should be cultivated with local scientists and public health officials. This would help researchers get a feel for the local ethical and political attitudes.

These partnerships would be part of a broader infrastructure needed to appropriately integrate local data collection with large-scale genomic science. To carry out what the group called "big science," consortia would be needed involving partnerships between public and private institutions, to help address multiple diseases and to advance funding opportunities, and between scientists and legal professionals, to aid in obtaining informed consent and addressing privacy issues.

Finding pathways and selecting targets: Quite an obstacle course

Inferring pathways of host-pathogen interaction from the data on genomic variation would be a daunting task. The group decided protein-protein interactions would be the place to start and called for a systems biology approach, in which many complex interactions are integrated in order to produce a model of the whole system. They also acknowledged the need for computational tools to make inferences about interactions. Statistically speaking, dealing with three sources of variation—in the host, the pathogen, and the environment—would be a challenge.

Once the pathways are found, possible therapeutic or vaccine targets would have to be identified. This is a challenge in drug or vaccine development because target selection has been notoriously poor in the past, leading to failed product development. "Knowing the variation of the host and pathogen allows us to select targets that are less likely to fail," one member assured the group. Once the candidate gene is identified through genome sequencing, the key would be to select multiple targets, looking at other genes found near the candidate gene that may influence its function. With the targets in hand, the next step would be to carry out standard methods of drug development, including target validation in animal models, functional assay development, and high-throughput screening to identify promising leads. The group decided the main obstacle for this phase of the project would be developing appropriate assays, which are expensive and time consuming. One solution would be to bring in additional funding from private companies at this stage. The group said these funding sources usually come in at a later stage when there is less risk, but involving them earlier would be crucial for this project.

Societal benefits: Where health goes, money follows

"If you improve health in general in a population, you get what's called a demographic transition," a group member said. After an initial boom in population, people start choosing to have smaller families, which helps lead to economic development. In addition to the obvious benefits of improving health and economics, these studies would increase research and development capacity in developing countries by building facilities and forming partnerships between local scientists and large institutions.

From a scientific standpoint, the group felt the project would increase basic understanding of how humans and pathogens interact, and provide a model for interdisciplinary science.

There was some discussion of the possible advances in personalized medicine that could come from these studies. The group agreed that a "one size fits all" approach is the norm right now in drug development and that this approach has serious limitations. One group member described the process of prescribing medicine as a game of trial and error. The doctor says, "Try this, it usually works."

The studies that the group proposed could be geared toward improving the "one size fits all" method by looking for a "magic bullet" that would really fit all. But the group members agreed that finding magic bullet cures for diseases is very unlikely and decided the studies should be designed to encourage development of more precise personalized medicine, where the right therapy is selected for the right person based on the person's genotype.

By the end of the four sessions most of the group members were satisfied with the plan they had proposed. They called for science at its grandest: large cohort studies built on complex infrastructures and partnerships with the goal of finding therapies for the world's deadliest diseases, using natural genomic variation as a guide. Some thought the strategy should involve more population biology, and others weren't sure whether the question was answered as fully as it could have been, but most were content.

"It was kind of nerve wracking at first, but it came together," one group member mused. As the group got up to leave at the end of the last session, another member said, "So we're done. It's kind of sad. It's like the end of summer camp."

WORKING GROUP SUMMARY – GROUP 2

Summary written by:

Chandra Shekhar, Graduate Student, Science Writing, University of California, Santa Cruz

Working group members:

- Agnes Awomoyi, Microbiology and Immunology, University of Maryland, Baltimore
- Phillip Berman, Scientific Director, Global Solutions for Infectious Diseases
- Bruce Beutler, Professor, Department of Immunology, The Scripps Research Institute
- Karen T. Cuenco, Research Assistant Professor, School of Medicine, Boston University
- Dennis Drayna, Chief, Section on Systems Biology of Communication Disorders, National Institute on Deafness and Other Communication Disorders, National Institutes of Health
- Michael Fasullo, Senior Research Scientist, Cancer Research, Ordway Research Institute
- Jonathan Kahn, Assistant Professor, School of Law, Hamline University
- Rob Knight, Assistant Professor, Chemistry and Biochemistry, University of Colorado, Boulder
- Erin McClelland, Research Fellow, Medicine, Albert Einstein College of Medicine
- Bob Roehr, Freelance Science Writer
- Michael Rose, Director, Intercampus Research Program on Experimental Evolution, University of California Systemwide
- Chandra Shekhar, Graduate Student, Science Writing, University of California, Santa Cruz
- Hongmin Sun, Life Sciences Institute, University of Michigan
- Shan Wang, Associate Professor, Materials Science and Engineering, Stanford University

Summary

"TB is a very attractive disease."

Taken out of context the above remark may seem bizarre, but spoken during a focus group discussion on natural variation in disease resistance it made perfect sense. Group members wanted to start by picking a well-known disease that clearly affected some people while leaving others in the same community unaffected. Based on the genomic principles that underlie this variation, the group would propose research projects to develop new therapies for infectious disease.

The first step was agreeing on a model disease. One member proposed TB. Another preferred sepsis. Others suggested flu, HIV, malaria, and smallpox. "Looks like everyone is trying to push their favorite disease," one member commented.

An hour of spirited debate followed. The TB camp battled the HIV camp. The bacterium faction crossed swords with the virus faction, as the plasmodium fans watched with amusement. The sepsis camp tried to enter the fray but was quickly routed by the chronic disease contingent. In the midst of this intellectual sparring, some members felt we should not focus on one disease but on common disease pathways.

The group finally hammered out a consensus: we would examine the effect of genomic variations on disease outcomes—genotype-phenotype relationships—for high-impact infectious diseases, including poorly understood ones. Some diseases, like malaria, have well-known polymorphisms, with different genes leading to different outcomes. Do similar polymorphisms exist for other diseases?

A visitor who joined the group's second session pointed out the benefits of using mouse models to identify mechanisms underlying such polymorphisms. It is relatively easy to manipulate a mouse's genome by knocking out or modifying genes. It then becomes possible to work backward from the phenotype: selecting disease-resistant specimens from randomly mutated mice and identifying the relevant gene mutations. These mutant mice can be quickly bred in vast numbers to study genetic aspects of disease resistance.

Human genetic studies, in contrast, are much more limited in scope: because humans can't be genetically manipulated, one has to work with a relatively smaller pool of naturally existing variations.

Although mouse models can yield useful results, as one member pointed out, "mice are not humans." Humans and mice share 98 percent of their genes and have similar disease pathways. But mice have a few specialized genes with no human counterparts. So one has to be cautious in applying insights from mouse models—or from any other animal models—to human diseases. Nonetheless, studying the effects of mutated mouse genes is a powerful technique for linking genetic variations to disease resistance. It is fast, versatile, and relatively safe.

The group had concluded the previous session with the idea of finding high-impact diseases with significant polymorphisms. As we continued the discussion, we decided to distinguish between initial infection and disease outcome. Some hosts are more resistant to infection but once infected, succumb rapidly. Others catch infections easily but tolerate them better. Yet others resist both infection and spread of disease or are vulnerable to both. Different polymorphisms expressed at the onset and during the progress of disease may be responsible for these variations.

One member mentioned a genomic chip that researchers are developing to track a host's responses to disease. Such a chip could shed light on such genotype-phenotype relationships; for instance, it could measure the level of cytokines, or proteins, the body uses to fight viral infection.

The group's third session began with a discussion of the wide gap separating the science and the clinical applications of genomics. While we came up with ideas for exploration in both science and applications, we saw our task as bridging the gap between them.

Genomics is undoubtedly a powerful tool, but pure genomic information is only a beginning. A bicycle is much more than an inventory of part diagrams. Likewise, an organism is more than an inventory of gene sequences. Existing genomic databases tend to be sparsely annotated—information about what the genes do is often absent, or described elsewhere in the literature. To make genomics useful, it is necessary to relate genes to their functions and describe the physiological interactions between them. In other words, genomics must make full use of old-fashioned genetics, gene by gene.

The human genome doesn't exist in isolation. The genome's medical environment includes the health and nutritional status of the patient, the drugs administered by physicians, even the complex of microbes that live on the skin, in the digestive or reproductive tracts, or elsewhere in the body. Because this environment influences how each human gene is expressed, it should be considered when describing the gene's function. Mouse models can help in establishing such linkages between gene expression and environmental variations.

The group went on to discuss the state of the art in clinical genomics. Tools now exist to model variation in patient and pathogen genomes. It is possible to model the patient's immune responses, both innate and acquired, at any given instant, although relating them to gene expression could still be a challenge. Existing tools can also provide a genomic snapshot of co-infecting pathogens. The group coined the term "static characterization" to describe current capabilities, because they do not yet permit us to determine the dynamics of infection and immune response.

The group went on to discuss the time-varying and evolving aspects of infectious diseases.

Some diseases, such as HIV and TB, can be either acute or chronic. What genetic variation could trigger a switch between them? In TB the transition is signaled by the activation of macrophages—cells produced by the immune system. Is it possible to find similar pathways for other diseases? Would this help in devising therapies?

How do pathogens evolve inside a host? Is it different for acute and chronic disease? Is it affected by the host response? How does the environment influence this? When several pathogens are present, how do they interact? And what is the role played by commensals—microbes normally present in the body?

As the ideas flowed, a pattern emerged. Dynamic characterization of infectious disease—tracking the pathogen genome as well as the host immune response in real time—seemed key to understanding the genomics of disease resistance and devising new therapies. It would put several new and powerful tools in the hands of physicians. It would allow them to monitor the patient's immune responses, both innate and acquired, at any given instant, much as they now dynamically characterize blood pressure or serum oxygen levels in hospitals. It would permit them to perform population genetic assays of the pathogens infecting the patient. It would also eventually enable them to track the patient's physiological genomics, as different genes are expressed and the immune system adjusts its response to fight the evolving infection.

At the final session the group worked on refining and tightening its research ideas into two main proposals; each would be a bridge between genomic science and clinical genomics.

The first bridge addresses an evolutionary question: can we predict pathogen evolution as a function of host genome and environmental variations? Attempts to predict the evolution of influenza virus have already been published in the literature; our group was interested in seeing whether this was a practical goal for biomedical genomics generally. Each host is essentially a new evolutionary experiment for the pathogen. The progress of an infectious disease depends on how well the pathogen adapts to its environment. Our group proposed to track the ensemble of pathogen genomes as they adapt to the host under a range of environmental variations, including presence or absence of therapeutic intervention. We proposed to test this idea using model systems, such as bacteriophages in different types of bacterial host. With a bacterial model system, this should be a feasible short-term research project; the work could later be extended to mouse models.

The second bridge deals with an ecological question: can we understand how co-infecting pathogens affect one another during the disease process? Certain co-infections, such as HIV and TB, are synergistic, whereas others, such as hepatitis B and C, are antagonistic. Why this happens is unclear, but genomic factors, both in the host and pathogen, almost certainly play a role. Solving this puzzle, we believe, could yield fundamental insights into host-pathogen interaction in successful resistance to infection. Again, the group proposed testing the importance of this phenomenon using a model systems approach: multiple phage infections of bacteria.

These two genomic bridges will show whether the evolutionary and ecological dynamics of infection are tractable and causally important (as we believe them to be). If this is indeed the case, it will establish the value of dynamic characterizations of disease genomics, leading to new therapies dynamically tailored to disease course. This will be an important advance over current medical treatment of infectious disease.

Appendixes

The 3rd Annual National Academies Keck *Futures Initiative* Conference

The Genomic Revolution:
Implications for Treatment and Control of Infectious Disease
Arnold and Mabel Beckman Center, Irvine, California
November 10-13, 2005

AGENDA

Wednesday, November 9 (Hyatt Regency Newport Beach)

6:00 – 10:00 p.m.	Welcome Reception / Registration – Garden Room 1 & Garden

Thursday, November 10 (Arnold and Mabel Beckman Center of the National Academies)

7:45 and 8:15 a.m.	Bus pick-up from the Hyatt Regency Newport Beach to the Beckman Center
8:00 a.m.	Registration (Outside Auditorium)
8:00 – 9:00 a.m.	Breakfast (Dining Room)
9:00 – 9:30 a.m.	Welcome and Opening Remarks (Auditorium) Wm. A. Wulf, President, National Academy of Engineering Harvey V. Fineberg, President, Institute of Medicine Richard N. Foster, Board Member, W.M. Keck Foundation Robert Waterston, Chair, Genomics Steering Committee

9:30 – 10:30 a.m.	Overview "Tutorial" Sessions *Genomics, Structural Biology, and Rational Vaccine Design* Gary J. Nabel Director of the Vaccine Research Center National Institute of Allergy and Infectious Diseases National Institutes of Health *Diversity of Human Microbial Pathogens and Commensals / Host-Pathogen Interaction (Part I)* David Relman Associate Professor of Microbiology & Immunology and of Medicine Stanford University Chief, Infectious Diseases Section Veterans Administration Palo Alto Health Care System *Question and Answer section for these two presentations will take place at 11:30*
10:30 – 11:00 a.m.	Break (Atrium)
11:00 a.m. – 12:15 p.m.	Overview "Tutorial" Sessions / Q&A *Diversity of Human Microbial Pathogens and Commensals / Host-Pathogen Interaction (Part II)* David Relman *11:30 – 12:15 Q&A – Gary Nabel, David Relman, and Claire Fraser*
12:15 – 1:30 p.m.	Lunch (Dining Room)

1:30 – 3:00 p.m. Overview "Tutorial" Sessions / Q&A

Team Science
Mary E. Lidstrom
Vice Provost of Research
Professor in Chemical Engineering and Microbiology
Frank Jungers Chair of Engineering
University of Washington

Team Science: The Microscale Life Sciences Center (MLSC)
Deirdre Meldrum
Director, NIH CEGS Microscale Life Sciences Center and the UW Genomation Laboratory
Professor of Electrical Engineering
University of Washington

2:30 – 3:00 Q&A

3:00 – 3:30 p.m. Task to Working Group

3:30 – 4:00 p.m. Break (Atrium / Palm Court 2 / Bay View 2)

4:00 – 6:00 p.m. Working Group Session 1 (Locations throughout Beckman Center)

2.	Technology to improve rapid response.	Bay View II – 2nd floor
3.	Develop an inexpensive diagnostic test.	Laguna – 2nd floor
5.	Spend $100 million to prevent the next pandemic flu.	Emerald Bay – 2nd floor
6.	Can genomics facilitate vaccine development?	Irvine Cove – 2nd floor
9.	Develop a device to detect and identify pathogens.	Board Room – 1st floor

10.	Shared pathways of attack for prevention.	Harbour – 2nd floor
11-1.	Role of public health in integrating genomics.	Balboa – 1st floor
11-2.	Role of public health in integrating genomics.	Newport – 1st floor
12.	Sequence an individual's genome for under $1,000.	Crystal Cove – 1st floor
14-1.	Natural variation in disease resistance.	Lido – 2nd floor
14-2.	Natural variation in disease resistance.	Back Bay – 2nd floor

6:00 – 7:00 p.m.	Reception / Networking
7:00 – 9:00 p.m.	Dinner and Communication Awards Presentation (Atrium)
9:00 p.m.	Buses depart Beckman Center for Hyatt Regency Newport Beach
9:00 – 11:00 p.m.	Informal Discussions / Hospitality Room Hyatt Regency Newport Beach – Garden Room 1 and Garden

Friday, November 11 (Beckman Center)

7:15 and 7:45 a.m.	Bus pick-up from the Hyatt Regency Newport Beach to the Beckman Center
7:30 – 8:30 a.m.	Breakfast (Dining Room)
8:30 – 10:45 a.m.	Overview "Tutorial" Sessions / Q&A (Auditorium)

Some Roles of Computation in Molecular Biology
Michael Waterman
University Professor
Professor of Biological Sciences, Mathematics, and Computer Science
University of Southern California

Human Genetic Variation
Leonid Kruglyak
Professor of Ecology and Evolutionary Biology and the Lewis-Sigler Institute for Integrative Genomics
Princeton University

Dual Meaning of Dual Use
Robert Cook-Deegan
Director, Center for Genome Ethics, Law, and Policy
Institute for Genome Sciences and Policy
Duke University

10:00 – 10:45 Q&A

10:45 – 11:15 a.m. Break (Atrium) / Friday Poster Set-up

11:15 a.m. – 12:45 p.m. Overview "Tutorial" Sessions / Q&A

Issues from Developing Countries: What are their needs? What are their unique delivery and implementation issues (access, cost, power requirements, transportability, etc.)?
Austin Demby
Senior Staff Fellow
Global AIDS Program
Centers for Disease Control and Prevention

	Genetic Analysis of Innate Immune Sensing Bruce Beutler Professor Department of Immunology Scripps Research Institute *12:15 – 12:45 Q&A*
12:45 – 2:00 p.m.	Lunch (Friday session posters available for previewing)
2:00 – 3:45 p.m.	Working Group Session 2 (Same meeting places as session 1) *(3:00 – 3:45 — Coffee and refreshments will be available in the Atrium, Palm Court 2, and Bay View 2)*
3:45 – 5:00 p.m.	Working Group Report Outs (Each group gives a 5 minute debrief) (Aud.)
5:00 – 6:30 p.m.	Friday Poster Session 5:00 – 5:45 p.m. Odd numbered posters are attended 5:45 – 6:30 p.m. Even numbered posters are attended (Refreshments will be served Atrium)
6:30 p.m.	Buses depart Beckman Center for Hyatt Regency Newport Beach
7:00 – 9:00 p.m.	Buffet Dinner – Hyatt Regency Newport Beach – Terrace Room
9:00 – 11:00 p.m.	Informal Discussions / Hospitality Room Hyatt Regency Newport Beach – Garden Room 1 and Garden

Saturday, November 12 (Beckman Center)

7:45 and 8:15 a.m.	Bus pick-up from the Hyatt Regency Newport Beach to the Beckman Center
8:00 – 9:00 a.m.	Breakfast (Dining Room)
9:00 – 10:30 a.m.	Overview "Tutorial" Sessions / Q&A (Auditorium) *Human Genome Sequencing at $5,000 a Pop* Robert H. Waterston Head, Department of Genome Sciences William H. Gates III Chair of Biomedical Sciences University of Washington School of Medicine *Microsystems* Todd Thorsen Assistant Professor of Mechanical Engineering Massachusetts Institute of Technology *10:00 – 10:30 Q&A*
10:30 – 11:00 a.m.	Break (Atrium) / Saturday Poster Set-up
11:00 a.m. – 1:00 p.m.	Working Group Session 3 (Same meeting places as session 1)
1:00 – 2:00 p.m.	Lunch (Saturday session posters available for previewing)
2:00 – 3:30 p.m.	Saturday Poster Session 2:00 – 2:45 p.m. Odd numbered posters are attended 2:45 – 3:30 p.m. Even numbered posters are attended *(3:00 – 4:00 — Coffee and refreshments will be available in the Huntington Room, Palm Court 2, and Bay View 2)*

3:30 – 5:30 p.m.	Working Group Session 4 (Same meeting places as session 1)
5:30 – 6:30 p.m.	Networking / Reception
6:30 - 8:00 p.m.	Dinner (Atrium)
8:00 p.m.	Buses depart Beckman Center for Hyatt Regency Newport Beach
9:00 – 11:00 p.m.	Informal Discussions / Hospitality Room Hyatt Regency Newport Beach – Garden Room 1 and Garden

Sunday, November 13 (Beckman Center)

7:15 and 7:45 a.m.	Bus pick-up from the Hyatt Regency Newport Beach to the Beckman Center
7:30 – 8:30 a.m.	Breakfast (Dining Room)
8:30 – 10:15 a.m.	Working Group Report-Outs (Auditorium) (15 minutes per group)
10:15 – 10:45 a.m.	Break (Atrium)
10:45 a.m. – 12:00 p.m.	Working Group Report-Outs – continued (Auditorium)
12:00 – 1:00 p.m.	Lunch
12:00 and 1:00 p.m.	Buses depart for Hyatt Regency Newport Beach and John Wayne Airport

THE GENOMIC REVOLUTION: IMPLICATIONS FOR TREATMENT AND CONTROL OF INFECTIOUS DISEASE
WORKING GROUP TOPICS

TECHNOLOGY

1. Design a point-of-care diagnostics for rapid detection of viral and bacterial pathogens. (*not running*)

2. Identify what technological advances in the fields of science and engineering need to be developed (either new technology or novel integration of existing technologies) to improve rapid response to new or emerging diseases? For example, can carefully reengineered viruses or bacteria become the next generation of therapeutic agents? How can computational biology better integrate the vast amounts of genomic knowledge to assist these efforts?

3. Develop an inexpensive (and cost-effective) diagnostic test (infection, genotype) that could be deployed in countries with little scientific research infrastructure. How can nanotechnology and new rapid diagnostic methods for other targets be adapted to diagnose malaria species, drug-resistance mutations, and vaccine-resistance polymorphisms in malaria-endemic countries?

4. Can genetically modified organisms be used to control disease? (*not running*)

VACCINES / GENOMIC ANALYSIS AND SYNTHESIS

5. How would you spend $100 million over the next five years to prevent the next pandemic flu? What would be the research strategy to utilize fully the genomic sequences of the hosts and pathogens to accelerate the development of therapeutics and vaccines for its prevention and control?

6. How can genomics facilitate vaccine development? Would efficient methods of synthesis of genomes help? Can genomics help to improve the assessment of efficacy of vaccines? What are the safety concerns?

7. How can genomic analysis of immune evasion strategies facilitate vaccination against HIV? (*not running*)

8. How can genomic analysis of immune evasion strategies facilitate vaccination against malaria? (*not running*)

DIAGNOSIS

9. Develop a device to rapidly and sensitively detect and identify pathogens in an environment/population, spread either naturally or through deliberate acts. Can genomics help differentiate between natural and deliberate disease outbreak and provide evidence for attribution? Do we know what defines—both genotypically and phenotypically—a pathogen vs. a nonpathogenic invader? If not, how can we determine this?

10. Are there shared pathways of attack that might provide new avenues of prevention? How can we find them? Once identified, what methods can be developed to stop them?

11. Explore the emerging role of public health in integrating genomics in surveillance, outbreak investigations, and control and prevention of infectious diseases. (*two sections will be running due to high interest in this topic*)

NATURAL VARIATION

12. What will it take to sequence an individual person's genome for under $1,000 in ten years?

13. Can evolutionary models of the emergence of resistance in a pathogen combined with combinatorial treatment schemes be used to develop a strategy to hold the pathogen in check? (*not running*)

14. How can we use natural variation in disease resistance to understand host-pathogen interactions and devise new therapies? (*two sections will be running due to high interest in this topic*)

Participants

The National Academies Keck *Futures Initiative*
The Genomic Revolution:
Implications for Treatment and Control of Infectious Disease

Conference

Arnold and Mabel Beckman Center of the National Academies
(Irvine, CA)
November 10-13, 2005

Asem Alkhateeb
Postdoctoral Scholar
Department of Human Genetics
University of Chicago

Megan Atkinson
Senior Program Specialist
National Academies Keck *Futures Initiative*

Philip Awadalla
Assistant Professor
Department of Genetics
North Carolina State University

Agnes Awomoyi
Postdoctoral Scholar
Department of Microbiology and Immunology
University of Maryland, Baltimore

Myles Axton
Editor
Nature Genetics

John Barry
Writer

Benjamin Bates
Assistant Professor
School of Communication Studies
Ohio University

Phillip Berman
Scientific Director
Global Solutions for Infectious Diseases

Bruce Beutler
Professor
Department of Immunology
The Scripps Research Institute

Corey Binns
Graduate Student
Department of Science Journalism
New York University

Sally Blower
Professor
Department of Biomathematics
University of California, Los Angeles

Katie Brenner
Doctoral Candidate
Department of Bioengineering
California Institute of Technology

Steven Brenner
Associate Professor
Department of Plant & Microbial Biology
University of California, Berkeley

James Brody
Assistant Professor
Department of Biomedical Engineering
University of California, Irvine

Lawrence Brody
Senior Investigator
Genome Technology Branch
National Human Genome Research Institute

William Bunney, Jr.
Distinguished Professor
Della Martin Chair of Psychiatry
Department of Psychiatry and Human Behavior
University of California, Irvine

Karen Burg
Hunter Chair and Professor
Department of Bioengineering
Clemson University

Frederic Bushman
Professor
Department of Microbiology
University of Pennsylvania

Robert Carlson
Senior Scientist
Department of Electrical Engineering
University of Washington

Shu Chien
University Professor of Bioengineering and Medicine
Chair, Department of Bioengineering
University of California, San Diego

Ralph J. Cicerone
President
National Academy of Sciences

Alex Cohen
Research Associate / Programmer
National Academies Keck *Futures Initiative*

Gareth Cook
Repoter
The Boston Globe

Robert Cook-Deegan
Director, Center for Genome Ethics, Law and Policy
Institute for Genome Sciences and Policy
Duke University

Brad Cookson
Associate Professor
Department of Laboratory Medicine and Microbiology
University of Washington

Elliott Crouser
Assistant Professor
Division of Pulmonary and Critical Care Medicine
Ohio State University Medical Center

Karen T. Cuenco
Research Assistant Professor
School of Medicine
Boston University

Barbara Culliton
Deputy Editor
Health Affairs

Mary Jane Cunningham
Associate Director
Life Sciences and Health Group
Houston Advanced Research Center

Dat Dao
Director
Life Sciences and Health Group
Houston Advanced Research Center

Ronald W. Davis
Professor of Biochemistry and Genetics
Director, Stanford Genome Technology Center
Stanford University School of Medicine

Austin Demby
Senior Staff Fellow
Global AIDS Program
Centers for Disease Control and Prevention

George Dimopoulos
Assistant Professor
Department of Molecular Microbiology and Immunology
Johns Hopkins School of Public Health

Cecilia Dobbs
Graduate Student
Department of Science Journalism
New York University

Siobhan Dolan
Department of Obstetrics, Gynecology and Women's Health
Albert Einstein College of Medicine and March of Dimes Birth Defects Foundation

Dennis Drayna
Chief, Section on Systems Biology of Communication Disorders
National Institute on Deafness and Other Communication Disorders
National Institutes of Health

Georgia Dunston
Professor
Department of Microbiology
National Human Genome Center
Howard University

Leah Moore Eisenstadt
Graduate Student
Department of Science Journalism
Boston University

Alison Farrell
Senior Editor
Nature Medicine

Michael Fasullo
Senior Research Scientist
Department of Cancer Research
Ordway Research Institute

Jeffrey Feder
Associate Professor
Department of Biological Sciences
University of Notre Dame

Harvey V. Fineberg
President
Institute of Medicine

Daniel A. Fletcher
Assistant Professor
Department of Bioengineering
University of California, Berkeley

Roxanne Ford
Program Director
W.M. Keck Foundation

Richard N. Foster
Managing Partner
Foster Health Partners, LLC
Board Member
W.M. Keck Foundation

Stephanie Malia Fullerton
Assistant Professor
Department of Medical History and Ethics
University of Washington School of Medicine

Kenneth R. Fulton
Executive Director
National Academy of Sciences

Sonja Gerrard
Assistant Professor
Department of Epidemiology
University of Michigan

Peggy Girshman
Assistant Managing Editor
National Public Radio

Lenette Golding
Doctoral Student
Grady College of Journalism and Mass Communication
University of Georgia

Ananda Goldrath
Assistant Professor
Department of Biology
University of California, San Diego

Mark Guyer
Director
Division of Extramural Research
The National Human Genome Research Institute

Kiryn Haslinger
Science Writer
Cold Spring Harbor Laboratory

David Haussler
Director of the Center for Biomolecular Science and Engineering
Professor of Biomolecular Engineering
Howard Hughes Medical Institute
University of California, Santa Cruz

Lyla M. Hernandez
Senior Program Officer
Board on Health Sciences Policy
Institute of Medicine

Ezra C. Holston
Assistant Professor
School of Nursing
University of California, Los Angeles

Barbara R. Jasny
Supervisory Senior Editor
Science

Stephen Albert Johnston
Director and Professor
Center for Innovations in Medicine
Biodesign Institute at Arizona State University

Jonathan Kahn
Assistant Professor
School of Law
Hamline University

Amos Kenigsberg
Graduate Student
Center for Science and Medical Journalism
Boston University

Rima F. Khabbaz
Acting Deputy Director
National Center for Infectious Diseases

Muin J. Khoury
Director
Office of Genomics and Disease Prevention
Centers for Disease Control and Prevention

Rob Knight
Assistant Professor
Department of Chemistry and Biochemistry
University of Colorado, Boulder

Leonid Kruglyak
Professor of Ecology and Evolutionary Biology
Lewis-Sigler Institute for Integrative Genomics
Princeton University

Paul Laibinis
Professor
Department of Chemical Engineering
Vanderbilt University

Corinne Lengsfeld
Associate Professor
Department of Engineering
University of Denver

Kam Leong
Professor
Department of Biomedical Engineering
Johns Hopkins University

Rachel Lesinski
Senior Program Specialist
National Academies Keck *Futures Initiative*

Thomas Levenson
Associate Professor
Massachusetts Institute of Technology

Mary E. Lidstrom
Associate Dean
New Initiatives in Engineering
University of Washington

Hod Lipson
Assistant Professor
Department of Mechanical and Aerospace Engineering
Cornell University

Robin Liu
Manager
Microfluidics Biochip
Combimatrix Corporation

Michael Lorenz
Assistant Professor
Department of Microbiology and Molecular Genetics
University of Texas Health Science Center

Allison Loudermilk
Graduate Student
Grady College of Journalism and Mass Communication
University of Georgia

Dan Luo
Assistant Professor
Department of Biological and Environmental Engineering
Cornell University

Alan McBride
Researcher
Goncalo Moniz Research Center
Oswaldo Cruz Foundation

Colleen McBride
Chief
Social and Behavioral Research Branch
National Human Genome Research Institute

Catherine McCarty
Interim Director and Senior Research Scientist
Center for Human Genetics
Marshfield Clinic Research Foundation

Erin McClelland
Research Fellow
Department of Medicine
Albert Einstein College of Medicine

Susanne McDowell
Graduate Student
Department of Science Communication
University of California, Santa Cruz

Ulrich Melcher
R.J. Sirny Professor
Department of Biochemistry and Molecular Biology
Oklahoma State University

Deirdre Meldrum
Director of the NIH Center of Excellence in Genomic Science (CEGS) Microscale Life Sciences Center
Professor of Electrical Engineering
University of Washington

Arcady Mushegian
Director
Bioinformatics Center
Stowers Institute for Medical Research

Isaac Mwase
Associate Professor
Department of Philosophy and Bioethics
National Center for Bioethics in Research and Healthcare
Tuskegee University

Gary Nabel
Director
Vaccine Research Center
National Institute of Allergy and Infectious Disease
National Institutes of Health

Daniel Oerther
Associate Professor
Department of Civil and Environmental Engineering
University of Cincinnati

Susan Okie
Contributing Editor
New England Journal of Medicine

Marc Orbach
Associate Professor
Division of Plant Pathology
University of Arizona

George O'Toole
Associate Professor
Department of Microbiology and Immunology
Dartmouth Medical School

Mihri Ozkan
Assistant Professor
Department of Electrical Engineering
University of California, Riverside

Claire Panosian
Professor of Medicine
Division of Infectious Diseases
David Geffen School of Medicine at University of California, Los Angeles

Babak Parviz
Assistant Professor
Department of Electrical Engineering
University of Washington

Fabienne Paumet
Associate Research Scientist
Department of Physiology and Biophysics
Columbia University

Aria Pearson
Graduate Student
Science Writing
University of California, Santa Cruz

P. Hunter Peckham
Director, FES Center
Professor
Department of Biomedical Engineering
Case Western Reserve University

Marty Perreault
Program Director
National Academies Keck *Futures Initiative*

Gregory A. Petsko
Gyula and Katica Tauber Professor of Biochemistry and Chemistry
Director, Rosenstiel Center
Rosenstiel Basic Medical Sciences Research Center
Brandeis University

Christopher Plowe
Professor
Department of Medicine
University of Maryland School of Medicine

Haley Poland
Graduate Student
Annenberg School of Journalism
University of Southern California

Alan Porter
Evaluation Coordinating Consultant
National Academies Keck *Futures Initiative*

Mary Reichler
Centers for Disease Control and Prevention

David Relman
Assistant Professor
Division of Infectious Diseases
Stanford University

Karin Remington
Vice-President
Bioinformatics Research
J. Craig Venter Institute

Anne W. Rimoin
Assistant Professor
Department of Epidemiology
School of Public Health
University of California, Los Angeles

Bob Roehr
Freelance Science Writer

Dave Roessner
Senior Evaluation Consultant
National Academies Keck *Futures Initiative*

Michael Rose
Director
Intercampus Research Program on Experimental Evolution
University of California (Systemwide)

Charles Rotimi
Professor and Director
National Human Genome Center
Howard University

Bernhard Rupp
Structural Genomics Group Leader
Biosciences
University of California – Lawrence Livermore National Lab

Daniel Salsbury
Managing Editor
Proceedings of the National Academy of Sciences

Paul Schaudies
Assistant Vice-President
Biological and Chemical Defense Division
Science Applications International Corporation

Holger Schmidt
Assistant Professor
Department of Electrical Engineering
University of California, Santa Cruz

Debra Schwinn
James B. Duke Professor
Department of Anesthesiology and Pharmacology
Duke University Medical Center

Beatrice Seguin
Postdoctoral Fellow
Canadian Program on Genomics and Global Health
University of Toronto

Chandra Shekhar
Graduate Student
Department of Science Writing
University of California, Santa Cruz

Mona Singh
Assistant Professor
Computer Science and Lewis-Sigler Institute for Integrative Genomics
Princeton University

Upinder Singh
Assistant Professor
Department of Internal Medicine, Microbiology and Immunology
Stanford University

Bill Skane
Executive Director
Office of News & Public Information
The National Academies

Christina Smolke
Assistant Professor
Department of Chemical Engineering
California Institute of Technology

Katherine Spindler
Professor
Department of Microbiology and Immunology
University of Michigan Medical School

Jonathan Stroud
Graduate Student
Annenberg School of Journalism
University of Southern California

Lubert Stryer
Winzer Professor, Emeritus
Department of Neurobiology
Stanford University School of Medicine

Shankar Subramaniam
Professor
Department of Bioengineering
University of California, San Diego

Diane Sullenberger
Executive Editor
Proceedings of the National Academy of Sciences

William Sullivan
Professor
Department of Molecular, Cell and Developmental Biology
University of California, Santa Cruz

Hongmin Sun
Life Science Institute
University of Michigan

Mercedes Talley
Program Director
W.M. Keck Foundation

Todd Thorsen
Assistant Professor
Department of Mechanical Engineering
Massachusetts Institute of Technology

Sarah Tishkoff
Associate Professor
Department of Biology
University of Maryland

Eric Topol
Provost and Chief Academic Officer
Cleveland Clinic Foundation

Victor Ugaz
Assistant Professor
Department of Chemical Engineering
Texas A&M University

Timothy Umland
Department of Structural Biology
Hauptman-Woodward Medical Research Institute

Luis Villarreal
Professor
Department of Molecular Biology and Biochemistry
University of California, Irvine

Joseph Vockley
Laboratory Director
Life Sciences Division
Science Applications International Corporation

Shan Wang
Associate Professor
Department of Materials Science and Engineering
Stanford University

Michael Waterman
Professor
Department of Computational Biology
University of Southern California

Robert Waterston
William H. Gates III Endowed Chair in Biomedical Sciences, Chair and Professor
Department of Genome Sciences
University of Washington

Debra Weiner
Attending Physician
Emergency Medicine
Children's Hospital Boston
Assistant Professor of Pediatrics
Harvard Medical School

Lloyd Whitman
Head
Code 6177
The Surface Nanoscience and Sensor Technology Section
Naval Research Laboratory

John Wikswo
Gordon A. Cain University Professor
Vanderbilt Institute for Integrated Biosystems Research and Education
Vanderbilt University

John E. Wiktorowicz
Associate Professor
Department of Human Biological Chemistry and Genetics
University of Texas Medical Branch

Huntington Willard
Director and Professor
Institute for Genome Sciences and Policy
Duke University

Marc S. Williams
Director
Clinical Genetics Institute
Intermountain Health Care

Mary Wilson
Associate Clinical Professor of Medicine and Associate Professor of Population and International Health
Harvard School of Public Health
Harvard Medical School

Elizabeth Winzeler
Associate Professor
Department of Cell Biology
The Scripps Research Institute

Steven Wolinsky
Samuel J. Sackett Professor of Medicine
Northwestern University Feinberg School of Medicine

Wm. A. Wulf
President
National Academy of Engineering

Zhenhua Yang
Assistant Professor
Department of Epidemiology
University of Michigan, School of Public Health

Hsiang-Yu Yuan
Graduate Student
Institute of Biomedical Sciences
Academia Sinica

Lyna Zhang
National Center for Infectious Disease
Centers for Disease Control and Prevention

Hongyu Zhao
Ira V. Hiscock Associate Professor
Department of Public Health and Genetics
Yale University

Shaying Zhao
Assistant Professor
Department of Biochemistry and Molecular Biology, and Institute of Bioinformatics
University of Georgia

Michael Zwick
Assistant Professor
Department of Human Genetics
Emory University School Of Medicine